produsing theory
in a digital world
2.0

Digital Formations

Steve Jones
General Editor

Vol. 99

The Digital Formations series is part of the Peter Lang Media and Communication list.
Every volume is peer reviewed and meets
the highest quality standards for content and production.

PETER LANG
New York • Bern • Frankfurt • Berlin
Brussels • Vienna • Oxford • Warsaw

produsing theory in a digital world 2.0

the intersection of audiences and production in contemporary theory

VOLUME 2

edited by rebecca ann lind

PETER LANG
New York • Bern • Frankfurt • Berlin
Brussels • Vienna • Oxford • Warsaw

The Library of Congress has catalogued Volume I as follows:

Produsing theory in a digital world: the intersection of audiences
and production in contemporary theory / edited by Rebecca Ann Lind.
p. cm. — (Digital formations; v. 80)
Includes bibliographical references and index.
1. Mass media—Technological innovations. 2. Mass media and technology.
3. Mass media—Social aspects. 4. Media literacy. 5. Digital media. 6. Social media.
I. Lind, Rebecca Ann. II. Title: Producing theory in a digital world.
P96.T42P76   302.23—dc23   2012014097
ISBN 978-1-4331-1520-2 (Volume 1 hardcover)
ISBN 978-1-4331-1519-6 (Volume 1 paperback)
ISBN 978-1-4539-0840-2 (Volume 1 e-book)
ISBN 978-1-4331-2729-8 (Volume 2 hardcover)
ISBN 978-1-4331-2728-1 (Volume 2 paperback)
ISBN 978-1-4539-1629-2 (Volume 2 e-book)
ISSN 1526-3169

Bibliographic information published by **Die Deutsche Nationalbibliothek**.
Die Deutsche Nationalbibliothek lists this publication in the "Deutsche
Nationalbibliografie"; detailed bibliographic data are available
on the Internet at http://dnb.d-nb.de/.

Cover image: SONOMA I, by Chuck Sabec (2015).
Reprinted with kind permission of the artist and Studio Eight Fine Art
(www.StudioEightFineArt.com). All rights reserved.
Layout of interior text by ChiTownMuggle.

The paper in this book meets the guidelines for permanence and durability
of the Committee on Production Guidelines for Book Longevity
of the Council of Library Resources.

© 2015 Peter Lang Publishing, Inc., New York
29 Broadway, 18th floor, New York, NY 10006
www.peterlang.com

Printed in the United States of America

Contents

Acknowledgments.. vii

1 Produsing Theory in a Digital World: Life in the Interstices
   REBECCA ANN LIND 1

2 The Interpretive Community Redux:
   The Once and Future Saga of a Media Studies Concept
   THOMAS R. LINDLOF.. 19

3 Duality Squared: On Structuration of Internet Governance
   DMITRY EPSTEIN.. 41

4 Produsing the Hidden: Darknet Consummativities
   JEREMY HUNSINGER 57

5 Online Performative Identity Theory:
   A Preliminary Model for Social Media's
   Impact on Adolescent Identity Formation
   BRADLEY W. GORHAM AND JAIME R. RICCIO............................... 75

6 Understanding the Popularity of Social Media: Flow Theory,
   Optimal Experience, and Public Media Engagement
   JOHN V. PAVLIK 91

7 "For this much work, I need a Guild card!":
   Video Gameplay as a (Demanding) Coproduction
   NICHOLAS DAVID BOWMAN 107

8 The Mobile Conversion, Internet Regression, and the
   Repassification of the Media Audience
   PHILIP M. NAPOLI AND JONATHAN A. OBAR.............................. 125

9   Social Media Audience Metrics as a New Form of
      TV Audience Measurement

      DARRYL WOODFORD, BEN GOLDSMITH, AND AXEL BRUNS......... 141

10   Staging the Subaltern Self and the Subaltern Other:
      Digital Labor and Digital Leisure in ICT4D

      RADHIKA GAJJALA, DINAH TETTEH, AND ANCA BIRZESCU......... 159

11   Race, Gender, and Virtual Inequality: Exploring the
      Liberatory Potential of Black Cyberfeminist Theory

      KISHONNA L. GRAY 175

12   Digital Human Rights Reporting by Civilian Witnesses:
      Surmounting the Verification Barrier

      ELLA MCPHERSON.. 193

13   Twitter as a Pedagogical Tool in Higher Education

      RENEE HOBBS 211

14   Engaging Adolescents in Narrative Research and
      Interventions on Cyberbullying

      HEIDI VANDEBOSCH, PHILIPPE C. G. ADAM, KATH ALBURY,
      SARA BASTIAENSENS, JOHN DE WIT, STEPHANIE HEMELRYK
      DONALD, KATHLEEN VAN ROYEN, AND ANNE VERMEULEN........ 229

15   Produsing Ethics [for the Digital Near Future]

      ANNETTE N. MARKHAM 247

16   Afterword: What's So New About New Media?

      DENNIS K. DAVIS.. 267

      Contributors 279

      Index 285

Acknowledgments

My greatest thanks are to the contributors to this volume. It has been a pleasure working with them; besides sharing their intriguing ideas on these pages, they have been responsive, understanding, and willing to engage in some fairly intense conversations with me during the writing process. I hope they are pleased with the outcome.

As always, many thanks are due to my colleagues (especially Steve Jones and Zizi Papacharissi) in the Department of Communication at the University of Illinois at Chicago, and to the College of Liberal Arts and Sciences, for their support. I appreciate the always cheerful, prompt, and excellent work of research assistant Paul Couture.

Grateful acknowledgment is made to the following for permission to use copyrighted material:

To my parents and my students.

Produsing Theory in a Digital World: Life in the Interstices

Rebecca Ann Lind

The universe is made up of stories, not of atoms. (Rukeyser, 1968)

Without a doubt, our universe—digital and physical—is socially constructed. And if the universe is made up of stories, we can argue that increasingly (courtesy of new/digital/social media), these are stories of our own creation. Or can we? To what extent do our stories represent our unbridled expression and the full measure of our creativity? Are we free to create, share, and receive the stories we desire? Or are we inhibited in meaningful ways? In the introduction to the first Produsing Theory volume, I explored some of the tensions generated in the spaces enabled by the confluence of the formerly disparate activities of producing and consuming media (Lind, 2012). These tensions have not dissipated; our universe—digital and physical—although socially constructed, remains socially constricted.

This volume continues the exploration of the new worlds we inhabit, the interstitial spaces lying between freedom and control, between self and other, between exploration and inhibition, between the production and use of media. It provides a site at which varied theories—some still emerging—can intersect and shine a light into the spaces between what previously had been neatly separated and discrete components of media systems. In some settings, division by audience, content, and production settings remains useful (e.g., Lind, 2013), but this volume, like the first, is all about the interstices.

Each of the chapters in this book takes a different perspective in its approach to the spaces formed as a result of rapidly developing and swiftly deploying new communications technologies and social software. Each has some type of connection to what Axel Bruns (2008) called *produsage*, briefly defined as "the collaborative and continuous building and extending of existing content in pursuit of further improvement" (p. 21). Produsers, embodying the interstitial spaces they inhabit, enact a hybrid role in the system—they both produce and use media content.

In so doing, as is the case with all communicative acts, they are constructing their social realities. They create and recreate their social worlds, and in the process they may also join, strengthen, or perhaps abandon various communities. In the process, they may also explore and experiment with various identities. These ideas recur throughout this book, necessarily entwined with the tensions between structurally imposed limitations and human agency.

Creating Our (Constrained) Reality

In their pivotal book, Berger and Luckmann (1967) introduced the term *the social construction of reality* and provided a well-thought-out consideration of how our day-to-day and even mundane interactions generate/maintain shared mental representations. Reciprocal roles become habituated and eventually institutionalized. "Social order," they argued, "exists *only* as a product of human activity" (p. 52, emphasis in original). Both reality and knowledge, according to Berger and Luckmann, are socially relative and contextually situated. Through our interactions, subjective meanings become objective facticities, and human activities produce a world of things. We create institutions through engaging in processes of reciprocal habitualized activity, yet these institutions function as control mechanisms by "setting up predefined patterns of conduct" (p. 55). Even as we create our world, therefore, we create, recreate, and legitimate institutions, and we begin to see these socially constructed institutions—which restrict, limit, and inhibit us— as objective realities. Institutions are continually reified, the symbolic universes (belief systems) underlying them reinforced, and their supposed objective status augmented as they are passed on to subsequent generations. "They have always been there," we think, and we may not even be able to imagine a world without them. Often unaware, we continue to enact behaviors consistent with institutionally predetermined expectations and constraints. Nonnormative behaviors are at least discouraged and often punished; they are framed as deviant, not natural, outside the realm of what is (seen as) self-evident. These institutions limit us—yet we ourselves create and perpetuate them. To Berger and Luckmann, the relationship between structure and agency is a dialectical one.

Although Berger and Luckmann's consideration of the mutual influence of social structures and human interactions is important, it is not the only way scholars have approached the nature of the relationship between structure and agency. Bourdieu (1977), who also presumed a dialectical relationship between structure and agency, added to our understanding the ideas of habitus, field, and capital. Habitus "is an acquired system of generative schemes objectively adjusted to the particular conditions in which it is constituted," according to Bourdieu (p. 95); it "engenders all the thoughts, all the perceptions, and all the actions consistent with those conditions, and no others." Bourdieu's conceptualization of habitus bears some similarity to, but is different from, Berger and Luckmann's presentation of the social construction of reality. As Swartz (2012) noted, Bourdieu attended more fully to issues of class in socialization processes than Berger and Luckmann did, but Bourdieu did "not conceptualize the possibility of a 'deviant habitus'" (p. 110).

An extremely useful approach to exploring the interrelationships between structure and agency, fully applicable to the intersection of audience and production that is produsage, is provided in Dreier's (2008) theory of persons as situated participants in social practices. As I noted in the introduction to the first *Produsing Theory* volume, Dreier advocated "a conception of a structure of ongoing social practice in a set of linked and diverse, local social contexts" (p. 23). The contextualization of social practices is key, with "local" being a quasiphysical and sociomaterial construct. Although Dreier sees social structures and social practices as interrelated and mutually influencing, and social practices as producing and reproducing the social world (as do other theorists referenced here), he argued that "notions about an abstract, individual agency must be replaced by a contextual conception of personal modes of participation rendering personal abilities many-sided and variable" (p. 40).

To Dreier (2008), individuals act and interact within and across a variety of social contexts, with context defined as "a delineated, local place in social practice that is re-produced and changed by the linked activities of its participants and through its links with other places in a structure of social practice" (p. 23). Any given culture includes any number of social contexts, differing according to duration, degree of openness to participants, relative flexibility or restriction of the range of social practices deemed appropriate, definitiveness of structural arrangements, and so forth.

Of primary importance, in his focus on personal stability and change, Dreier (2008) rejected the pathologizing of the non-fully integrated or non-fully coherent personality. As a psychotherapist, he argued that for most people, varying one's social practices across contexts is adaptive behavior, not pathological. His work is concerned with "how personal stability and change are allowed and inhibited by a person's trajectory of participation in structures of social practice" (p. 40).

One of several benefits of considering the recursive relationship between structure and agency is that the very process of attending to these phenomena heightens our awareness and presumably the salience of forces that often remain hidden. If we acknowledge that we play an important role vis-à-vis socially constructed limitations, we may choose to do something about it. A number of contributors to this volume urge us to attend more actively to our roles in perpetuating or contesting hegemonic ideologies. The most emphatic challenge is issued by Markham (Chapter 15), who argues that in all of our actions, we create and recreate (or resist) socially constructed understandings of what is right, good, or just. "Ethics matter," she says, and her chapter is a clarion call for us to assume our responsibility to create the sort of world in which we would wish to live. Unless we do so, "the future will seem to just happen to us."

But if we wanted to move toward fostering a more acute and aware acknowledgment of the part we play in the social construction of reality, how might we think about engaging our agency?

Engaging Our Agency

Various theories include concepts relevant to the possibility that we may change our social worlds, including those explicitly concerned with structure and agency. For example, emancipatory potential is ever present in Giddens's (1984) structuration theory by virtue of his concept of reflexivity, which sees agents as holding the capability to act in such a way as to effect change (or, obviously, to perpetuate the status quo). Unger's (2001) antinecessitarian social theory presents the possibility that we might reject or transcend social rules. Social institutions can be "denaturalized" (pp. 125–126). Unger posited that our behaviors can reflect various types of agency and that empowerment can take multiple forms. Again, change is possible (although not without challenge in the face of existing structures); we are not necessarily doomed to repeat our past and reproduce what we have inherited.

Perhaps we might also increase the likelihood that we'll create the world we desire—and in that sense engage our agency—by approaching our social world in a more *intentional* manner. As a concept, intentionality appears with varying degrees of specificity and rigor across academic and professional fields as well as among the lay population. Presenting a folk-theoretical approach to intentionality, which is consistent with the point being made here, Malle, Moses, and Baldwin (2001) argued that

> those who endorse the folk-theoretical approach do not try to clarify the nature of explanation in the philosophical sense, nor do they necessarily try to postulate the objective existence of intentionality. Instead, they analyze explanations and intentionality as cognitive tools that guide people's perception, prediction, and control of their environment. (p. 17)

Malle et al. (2001) presented the concept of intentionality as a tool with a number of functions. This tool allows us to consider basic mental categories (e.g., desire, awareness) and to detect structure (intentions and actions) in human behavior. Perceivers determine others' intentionality by making inferences based on the behaviors, the context, and other characteristics. Such inferences are necessary because "the content of agents' intentions is radically underdetermined by their behavior" (p. 10), which otherwise could yield an almost infinite range of possibilities. According to Malle et al., the inference approach is consistent with constructivism, because our understandings are modified through social interactions. Intentionality "supports coordinated social interaction by helping people explain their own and others' behavior" and "plays a normative role in the social evaluation of behavior through its impact on assessments of responsibility and blame" (p. 1). Normative responsibility is defined as "a normative relation" (in the form of duties and liabilities) "between an agent and a specific action or outcome" (p. 20).

Such a view considers behaviors intentional if the explanations for the behaviors reflect the following five components: "a desire for that outcome; beliefs about an action that leads to that outcome; an intention to perform the action; skill to perform the action; and awareness of fulfilling the intention while performing the action" (Malle & Knobe, 1997, p. 111). It also ties responsibility to agency, or "the capacity to perform autonomous, rational action" (Malle et al., 2001, p. 22).

In contrast to folk-theoretical considerations of intentionality, philosopher Edmund Husserl's conception of intentionality is much broader, addressing any experience or object-directed thought. It was described by McIntyre and Smith (1989) as "a *phenomenological* property of mental states or experiences" (p. 150, emphasis in original). Husserl's articulations of intentionality include but go well beyond "the everyday notion of doing something 'intentionally': an action is intentional when done with . . . a mental state of 'aiming' toward a certain state of affairs" (p. 148) and thus include but go well beyond the folk-theoretical considerations of intentionality. Some of Husserl's thinking most relevant to the present discussion includes his considerations that we know about intentionality based on our own subjective experiences; we focus on what an act represents and how; and we connect our understanding of any given act to other mental states and experiences. He noted that as we encounter objects, we approach from a particular angle. The angle of approach necessarily provides a restricted and partial view of an object, rendering it at least partly indeterminate and making inevitable the possibility of having other experiences of/with that object. Each discrete perception of and experience with an object is but one of many possible experiences, each of which can be found on an open-ended *horizon* of possible perceptions and experiences. We may explore multiple perceptions of and experiences with an object, and if we do, we will attain a greater understanding of that object's properties. Objects have both internal and external horizons, differentiated by whether the properties are relational and which are constrained by our existing beliefs or presuppositions about the object or object type. The external horizon is particularly relevant to our considerations here; according to McIntyre and Smith, "the external horizon . . . reflects the fact that objects are not perceived as solitary things [but are] related to" other things (p. 174).

Even limiting our consideration of Husserl to these concepts—concepts that map onto a folk-theoretical understanding of intentionality—we can see the value of engaging with our social world in an intentional manner. We can actively consider how we will approach and respond to the situations we encounter. We can try to be more aware of the inferences we make about others and their intentionality. We can try harder to explore the limitations of the restricted and restrictive nature of our perceptions and experiences. We can try to be more aware of the behaviors we enact and the outcomes they are likely to produce. Again, although changing our social structures is challenging, if we strive to act with greater intentionality, we may enact the change we seek.

We might also choose to engage in *reflection* as we approach various situations and as we engage in the construction of the most desirable social world. The concepts of reflection and reflective practices have their roots in the work of John Dewey (1933, 1938) and Donald Schon (1983, 1987) and are fundamental in a number of fields. Indeed, as Farrell (2012) wrote, "it seems that the terms reflection and reflective practice are so popular in education that they are nearly mandatory" (p. 8) in teacher education and professional development.

Engaging in reflection has a clear connection to our objective of rising to the challenge of actively working to construct the world in which we wish to live. According to Dewey (1938), reflection "is the heart of intellectual organization and of the disciplined mind" (p. 10). As he conceived it, reflection is a holistic process leading to change and professional growth; reflective action requires open-mindedness (being open to multiple perspectives), responsibility (considering the consequences of actions), and wholeheartedness (continually examining assumptions, beliefs, and results and being open to learning). In contrast with reflective actions, routine actions are "guided primarily by impulse, tradition, and authority" (Zeichner & Liston, 2013, p. 10) and can "serve as a barrier to recognizing and experimenting with alternative viewpoints."

The process of inquiry during the reflective practice begins when we encounter an indeterminate situation and strive to clarify it or make it more determinate. Because the process of inquiry by definition engages us with the situation, Dewey believed that inquiry is social. However, Schon (1992)—whose "theory of reflective practice" (according to Kinsella, 2007, p. 103) "has gained unprecedented popularity in the professional discourses of the health and social sciences"—emphatically stated that even though Dewey acknowledged the social contexts in which transactions of inquiry occur, he was not a constructivist.

Schon's (1983) reworking of Dewey's conceptualization of reflective thought into what he called reflective practice similarly prioritized one's practical experiences but took an explicitly constructivist stance. He argued that professionals engage in a process of framing and reframing as a result of their experiences. Importantly, reflection includes not only problem solving but also problem setting, "a process in which, interactively, we name the things to which we will attend and frame the context in which we will attend to them" (p. 40). The process of solving problems and engaging in further reflections "spirals through stages of appreciation, action, and reappreciation. The unique and uncertain situation comes to be understood through the attempt to change it, and changed through the attempt to understand it" (p. 132).

Reflective practices may well be incorporated into attempts to actively construct a social world comprised of particular desired attributes. Attending not only to the nature of that to which we will respond (problem setting) but also to our specific responses (problem solving) does allow the possibility of change. If the reflective actor may engender change, then without reflection, change may be highly unlikely. Zeichner and Liston (2013) argued that unreflective actors "often lose sight of the purposes and ends toward which they are working and become merely the agents of others" (p. 10).

Kinsella (2012) expanded Schon's concept of reflective practice and linked it to *phronesis,* commonly presented as practical wisdom. Phronesis is one of Aristotle's intellectual virtues, and the practical wisdom with which it is concerned is often presented alongside techne (technical, productive knowledge) and episteme (theoretical, context-independent knowledge). Phronesis emphasizes reflection and action and is linked to morality. In her extension of Schon's work, Kinsella argued that Schon's presentation of reflectivity did not adequately address reflexivity. Further, although Schon's work is constructivist, Kinsella argued that it is not sufficiently socially constructivist: even though Schon attended to context, he did not "fully acknowledge the background and social conditions that implicitly influence and contribute to our ways of seeing" (p. 43). In correcting for these perceived shortcomings, Kinsella augmented Schon's concept to include reflexivity. She also extended his consideration of reflection, representing reflection as a series of types falling along a continuum. Her continuum incorporates Schon's intentional and embodied reflections, to which she added receptive (intuitive or contemplative) reflection.

Kinsella (2012) also explored a preliminary consideration of criteria for phronetic judgment in an attempt to foster phronesis. She complemented three criteria presented by Schon ([1987] pragmatic usefulness, persuasiveness, and aesthetic appeal) with three of her own (ethical imperatives, dialogic intersubjectivity, and transformative potential).

Capitalizing on our practical wisdom, especially that linked to reflective practices, ethics, and reflexivity, can afford valuable vantage points at which we can incorporate the virtue of phronesis and thus attempt to generate a desirable social world. The chapters in this volume provide glimpses into those and other vantage points. Intentionally invoking the suggestion made by Zizi Papacharissi in her Afterword to the first *Produsing Theory* volume—that we play with linearity—let us abandon numerical sequencing. Instead, what follows examines how the chapters make connections between the physical and digital worlds before turning to how they address unfolding perceptions and understandings of the audience and the various processes involved during produsage.

The Virtual Is Real

Markham (Chapter 15) argues that ethical issues permeate our rapidly evolving world and reinforce the value of phronesis in helping us consider the ramifications of our digital interactions. Because our communicative behaviors are necessarily part of our socially constructed understandings of right and wrong, we must actively engage with such understandings in the hopes of creating a better world. The intentionality Markham advocates is intertwined with a remix or "What if?" approach to ethics, and her analysis focuses on several points of intervention: the datafication of everything, online quizzes, research ethics as traditionally enacted by academic institutions, and public responses to ethical controversies. She makes a connection between the physical and virtual worlds—with a concomitant desire to create a more just and equitable social world for all inhabitants. Several other contributors to this volume also identify, to greater or lesser extent, the recursive relationship between the digital and the physical worlds.

Gray's (Chapter 11) consideration of the liberatory potential of Black cyberfeminist theory encourages us to "critically engage with the recursive relationship between our physical environments and our virtual selves" and, further, to "use the framework to improve women's lives." Gray presents Black cyberfeminism as an extension of cyberfeminism (criticized as utopic), technofeminism (insufficiently inclusive of all women's lived experiences), and Black feminist thought (which embraces activism and considers the unique standpoints of Black and other marginalized women). Gray argues that Black women have incorporated social media as tools with which to engage their ongoing physical-world struggles and that people of color are using virtual spaces to communicate among and empower themselves.

Similarly, McPherson (Chapter 12) considers the potential for our communicative behaviors in the virtual world to improve life in the physical world, specifically in helping organizations respond to human rights violations. In an interesting application of Bourdieu's concept of fields of production, she highlights the tension at the intersection of professional human rights fact finders (a field) and digital civilian witnesses (a non-field). The former, reliant on their institutional credibility, are both helped and hindered by the deluge of digital information provided by the latter—and which cannot be considered evidence of human rights violations unless verified. She discusses training and technological possibilities for enhancing the inclusion of verification strategies in civilian witnesses' submissions; these labors effectively serve as verification subsidies.

The intersection of the physical and the virtual worlds as approached by Gajjala, Tetteh and Birzescu (Chapter 10) focuses on the range of representations of what are called the digital subaltern 2.0 on the digital financial platforms Kiva.org (which provides microloans to aspiring entrepreneurs) and M-PESA (a mobile money system). The digital subaltern is often staged as Other, or perhaps transitioning into (but not quite there yet) an individualized Self, rather than as an individualized Self; users' digital avatars on Kiva are often framed in such a way that, despite having been prodused by the entrepreneurial borrowers, colonial hierarchies are reinscribed. Gajjala and colleagues argue that the "virtual enactments" on Kiva and M-PESA "are *real* in their material impact and consequence: they work to reconfigure access as access to a homogenized global financial and market space."

Vandebosch et al. (Chapter 14) investigate the phenomenon of cyberbullying and suggest that a bottom-up, qualitative approach will facilitate researchers' understanding of how cyberbullying can be responded to, detected, and prevented. They recommend collecting adolescents' own narratives of cyberbullying to complement the field's current focus on quantitative research and to inform effective novel interventions. They argue that young people can and should be involved at each stage of the processes of defining, detecting, and intervening in cyberbullying, and—in an analog to media literacy—that participating in anticyberbullying message production activities can in itself serve as a prevention strategy.

Hobbs (Chapter 13) also makes a connection between the physical and virtual worlds (and to media literacy). Her intent is to improve students' ongoing learning, and her hope is "that the online course experience builds knowledge and skills applicable to the world outside the classroom." In her essay, Hobbs describes her explorations with Twitter as a pedagogical tool, reflecting on its rich possibilities due in part to the fact that it represents an open learning environment. She believes that Twitter can be a powerful tool for learning but warns that educators must be aware of and responsive to the learning curve in using Twitter.

When considering how our communicative behaviors in (or about) the virtual world function to create and recreate our social systems, both Epstein (Chapter 3) and Hunsinger (Chapter 4) use a wide-angle lens to look at produsage.

Hunsinger contemplates darknets (hidden online information systems, contradistinctive to the consumer Internet) and their consummativities. In integrated world capitalism, consummativity is to consumption as productivity is to production; it is founded on a system of desires/needs that are culturally constructed, exploited, and realized (or not). Even as we produse mediated content (or, in this case, darknets as technologically enabled structures), we are cocreating their underlying systems of desires/needs. In darknets, privacy and security are prime consummativities, and although darknets exist for many reasons, the most common incentives center on surveillance, privacy, security, and economic value. Hunsinger notes that as produsers engage darknets and the knowledges contained therein, their awareness of their positions in regard to surveillant assemblages allows them to differentiate themselves from those who remain unaware; indeed, hiddenness becomes fetishized. Besides shining a light into these spaces, Hunsinger's focus on darknets—a subset of the larger Internet but one unfamiliar to many members of our society—allows us to more clearly understand the phenomenon of consummativity and highlights an important aspect of integrated world capitalism.

Epstein's (Chapter 3) work represents a structurational approach to understanding Internet governance. He argues that processes of governance are by nature conscious acts of social construction. He proposes a model that simultaneously addresses the structural aspects of policy and the relevant actors' human agency in shaping policy. Epstein's Duality Squared model addresses structures, social systems, and human agents in the Internet policy arena, allowing greater understanding of the multifaceted relationships between policy-making processes and outcomes. In this model, technology is both a medium and an outcome of human action; interactions with/via technologies both occur in and have consequences on structural conditions; policy is both a factor and an outcome of human activity; and structural conditions both influence and are influenced by policy. The model can help us understand a wide range of technology governance and policy-making domains, and as Epstein notes, "may be a particularly suitable framework for the study of bottom-up and multistakeholder processes such as Internet governance." It certainly helps us identify questions about policy-making processes and the relationships between such processes and relevant social structures.

The Audience I: Evolutions and Disruptions

The chapters by Napoli and Obar (Chapter 8), Lindlof (Chapter 2), and Woodford, Goldsmith, and Bruns (Chapter 9) interrogate the concept of the media audience—as an entity—in different ways. Of course, what we call "the audience" has long been of interest to scholars and practitioners. Is the audience active? Passive? Does it even exist? As a social construction, however, media audiences are alive and well and continue to warrant our attention, especially when homesteading the interstitial frontier.

Napoli and Obar discuss the evolution of media and their audiences, highlighting the extent to which audience activity and participation have been facilitated or discouraged over time. The authors wonder whether, contrary to the hope that the Internet would facilitate activity and autonomy, the ongoing mobile conversion represents an evolutionary regression, a step back from the Internet's liberatory potential. They argue that because of the different affordances of mobile and PC-based Internet technologies, coupled with the culture industry's predilection for an audience that is primarily seen as receivers of content, we may be undergoing a return to "more passive, consumption-oriented audiences." They describe differences in platform architectures (affording fewer opportunities for programming and innovation to mobile users) and differences in usage patterns (mobile users are more constrained in their information-seeking and content-creation behaviors).

Lindlof (Chapter 2) also notes the disruptive potentials of new technologies, affecting media industries as well as media scholars; the digital media revolution has had profound implications for what we consider "the audience" and requires that we reexamine our theoretical tools. Lindlof argues that the notion of the interpretive community remains a viable framework with which to study how people use and make sense of media. After describing the origins, proliferation, and limitations of the concept, he considers how it might evolve to best inform our understandings of a produsing audience. Besides the meaning-making processes inherent in reading an industrially produced text, scholars must attend to the additional meaning-making activities related to transmedia storytelling as audiences actively rewrite texts across any number of media platforms. We must attend to the digital media forum as a site of community; how do these forums function as communities, and how do such communities thrive in what we might call, in this context, the intertextual interstices? Lindlof challenges us to match the long tail of the digital economy with a "cultural long tail," or a "vision of incredibly numerous interpretive communities receding into the distance, each of them clinging to an aging text, updating the interpretive strategies applied to it, and sharing their passions online."

Digital media have contributed to significant disruptions in the business of audience measurement, as described by Woodford et al. (Chapter 9). Television broadcasters, especially in commercialized markets, have long relied on simple and authoritative measures of who is watching what. The authors argue that the demand for ratings data, as a common currency in transactions involving advertising and program content, will likely remain, but accompanying measurements of audience engagement with media content would also be of value. Today's media environment increasingly includes social media and second-screen use, providing a data trail that affords an opportunity to measure engagement. Woodford et al. acknowledge some of the limitations of using social media to indicate audience engagement but posit that if these can be overcome, social media use might allow for quantitative and qualitative measures of engagement. Such indicators might even represent "a postdemographic alternative to the audience segmentation models of conventional ratings." They argue that raw social media data must be contextualized and suggest incorporating tools used by sports analysts to do so. Inspired by baseball's Sabermetrics, the authors propose Telemetrics in an attempt to separate actual performance from contextual factors. Telemetrics facilitates measuring audience activity in a manner controlling for factors such as time slot, network, and so forth. It potentially allows both descriptive and predictive measures of engagement.

The Audience II: Produsing and the Black Box

Continuing our investigation of the audience in an evolving media environment, we turn to another set of chapters, which moves away from a focus on what we might call "audience as entity" (albeit an entity composed of individuals). These chapters look into the enigmatic black box of the human mind and apply psychological constructs to enhance our understanding of produsers.

Pavlik (Chapter 6) considers whether flow theory might help us comprehend our engagement with social media. Flow refers to a state of being highly focused and highly absorbed in an activity, such that one is completely immersed and time seems to slow. Pavlik argues that social media can readily satisfy the conditions associated with flow: variety, goal clarity, flexibility, immediate feedback, alignment of skill and challenge, connection to a larger whole, and an autotelic personality (one is rewarded by the activity itself rather than simply its results). He contrasts the qualities of social media with those of traditional media. Social media actively engage their users, provide multiple avenues of engagement that can align with greater or lesser skill levels, provide ongoing and often immediate feedback, and allow users to participate in any number of virtual communities. As such, social media are much more likely than traditional media to generate a flow state.

Video game players can certainly enter a flow state, but can the gaming experience become too demanding? Bowman (Chapter 7) highlights the multiple demands made by games on their players, who must function as coproducers of the texts, noting that demand might impinge on the players' enjoyment. Bowman argues that in making gameplay decisions (a requirement that is itself a hallmark of the medium), players become coauthors of their experiences. But these decisions are not always easy—in making them, players must tap into multiple cognitive, emotional, behavioral, and social capacities. Bowman reviews research associated with each of these types of demands and in so doing not only helps facilitate our understanding of players' engagement with games but also provides guidance regarding important areas for continued research.

Gorham and Riccio (Chapter 5) present a preliminary framework linking symbolic interactionism to the psychological effects of such interactions and begin to lay out the constructs relevant to a theory explaining the effects of ongoing collaborative identity performance (such as occurs via social media) on one's enduring schemas for self. Focusing on children and adolescents, Gorham and Riccio present social media use and social media feedback as independent variables, moderated by personality traits such as openness to experience, conscientiousness, extraversion, agreeableness, and neuroticism. Dependent variables include the strength of various schema for particular social identities. Ultimately, as the authors put it, "social media use becomes a cycle of identity performance, feedback, and reification as identity-congruent behaviors are rewarded and become more central to the self-schema." Gorham and Riccio's preliminary presentation of what they call online performative identity theory is worthy of continued development and attention, perhaps beginning with some of the fruitful avenues for research suggested by the authors themselves.

Reflections in/of/on the Interstices

A number of themes echo across the interstitial spaces created at the nexus of this *Produsing Theory* volume and the first.

Reflections on the media audience's relationship to an industry can be found, to different degrees, in this volume's contributions by Lindlof (Chapter 2), Napoli and Obar (Chapter 8), Woodford et al. (Chapter 9), and Gajjala et al. (Chapter 10), among others. Matt Hills's first-volume scrutiny of spoilers and fans' self-narratives, although rooted in psychoanalytic theory, clearly links fans to the media industry. Some contributors have looked at audience/industry intersections from a more macro-level perspective, in some cases considering capitalism or industrialization more broadly. These include Epstein (Chapter 3) and Hunsinger (Chapter 4) in this volume, as well as the chapters by Bolter, Booth, and Freedman in the first. Jay Bolter framed the increasing importance of procedurality in contemporary times as the latest stage of an advanced industrial society's mechanization. Paul Booth argued that the interactive, role-playing adventure game *MagiQuest* functions as a guide to living a constituted, capitalistic life. Eric Freedman theorized the construction of what he called the "life technobiographic" and (like Hunsinger, Chapter 4) acknowledged the link among organizations, research, technology, and subjectivity.

Whereas Booth's chapter (see also Bowman, Chapter 7) investigated actual game contexts, and Bolter included games among his examples, Freedman's work certainly facilitates a discussion of gamification and, correspondingly, what may be called datafication or dataization. A number of chapters in the present volume consider such issues, to a greater or lesser extent, including Pavlik (Chapter 6), Gajjala et al. (Chapter 10), Vandebosch et al. (Chapter 14), and Markham (Chapter 15).

However, although gamification can present ethical issues, sometimes it is called into service of education, awareness, or behavior change. In this way, digital and social media—gamified or not—can be framed as tools (presumably to be used in an intentional, reflective manner). Broadly considered, such tools may be applicable to any number of settings and designed to meet the interests of any one or more of the actors involved, as demonstrated in the chapters by Woodford et al. (Chapter 9), McPherson (Chapter 12), Hobbs (Chapter 13), and Vandebosch et al. (Chapter 14). Epstein's (Chapter 3) model—at least in a few of the relationships it captures—may also be consistent with such a perspective. The first-volume chapters by Bruns and Highfield, Lee and Webb, and Heilferty also at least implicitly acknowledge some payoffs of social media use, which could be approached as intentionally applicable tools. Bruns and Highfield, for example, studied citizen journalism and argued that Twitter turbocharges gatewatching practices, bringing us closer to what Lasica (2003) called "random acts of journalism." Lee and Webb's test of their identity, content, community (ICC) model of blog participation documented a link between blogging interaction and sense of community, which certainly could be seen as a goal in certain segments of the blogosphere. Heilferty's analysis of blogs created by parents of children with cancer found, among other themes, types of content that might be used actively to accomplish some desired objective (such as stress management and uncertainty management).

Finally, the two *Produsing Theory* volumes contain many reflections of how our representations of self may be inhibited by the very affordances of the technologies used to create them. These appear with various degrees of explicitness in this volume by Hunsinger (Chapter 4), Napoli and Obar (Chapter 8), Gajjala et al. (Chapter 10), and Markham (Chapter 15). They are also raised by Jay Bolter and Shayla Thiel-Stern in their contributions to the

first *Produsing Theory* volume—that we are becoming increasingly proce-
duralized and our communications increasingly templated.

As I argued in the first volume, what I call here the interstitial spaces
may be relatively new, but in them, we have reproduced many traditional
limitations. For example, Paul Booth's first-volume examination of
MagiQuest found that players reinforced traditional gender norms as well as
the normativity of Whiteness. Similarly, other chapters in the first volume
showed how even members of subordinated cultural groups produced,
reproduced, and prodused hegemonic cultural barriers: the analysis of the "It
Gets Better" project by Gust Yep, Miranda Olzman, and Allen Conkle; the
exploration of how amateur pornographers prodused the cuckold fantasy
(Diego Costa); and Shayla Thiel-Stern's study of young people's use of
social media to engage in public identity production.

Bolter's contribution to the first *Produsing Theory* volume made it clear
that because of the increasing importance of procedurality (in the form of
parameterization and event loops), although social media can and do allow
users to negotiate their identities, users cannot do so with complete freedom.
As we think about Bolter's argument in a digital media environment ever-
more reliant on mobile devices, what happens if Napoli and Obar (Chapter 8)
are right, not Gray (Chapter 11)? What if the trend toward mobile media
suppresses the Internet's radical potential?

Thus, digital media and other Web 2.0 technologies continue to offer
promise for enhancing our physical social worlds, but we have not reached
the proverbial Promised Land just yet. And we are unlikely to do so unless
we actively, intentionally, reflectively, perhaps even phronetically do so.

Still, engaging in produsage has great potential to reduce the power of
social structures and could even reshape democracy. Bruns (2008) posited:

> What may result from this renaissance of information, knowledge, and creative
> work, collaboratively developed, compiled, and shared under a produsage model,
> may be a fundamental reconfiguration of our cultural and intellectual life, and thus
> of society and democracy itself. (p. 34)

Taken together, these chapters illuminate some of the interstices that are
home to *Homo Irretitus,* described by Saulauskas (2000) as "the netting and
at the same time netted human being." The contributors shine multiple
spotlights into the intersection of audiences and production. They help guide
us on our continuing journey toward a nuanced understanding of the interstitial spaces. Yet, as with the first *Produsing Theory* volume, the chapters give
rise to a near-limitless array of questions. I hope you enjoy the process of
finding answers and uncovering questions as you read what follows.

References

Berger, P. L., & Luckmann, T. (1967). *The social construction of reality: A treatise in the sociology of knowledge.* New York, NY: Anchor Press.

Bourdieu, P. (1977). *Outline of a theory of practice.* Cambridge, UK: Cambridge University Press.

Bruns, A. (2008). *Blogs, Wikipedia, Second Life, and beyond: From production to produsage.* New York, NY: Peter Lang.

Dewey, J. (1933). *How we think.* Boston, MA: D. C. Heath.

Dewey, J. (1938). *Logic: The theory of inquiry.* New York, NY: Holt, Rinehart & Winston.

Dreier, O. (2008). *Psychotherapy in everyday life.* New York, NY: Cambridge University Press.

Farrell, T. S. C. (2012). Reflecting on reflective practice: (Re)visiting Dewey and Schon. *TESOL Journal, 3*(1), 7–16.

Giddens, A. (1984). *The constitution of society: Outline of the theory of structuration.* Berkeley: University of California Press.

Kinsella, E. A. (2007). Technical rationality in Schon's reflective practice: Dichotomous or non-dualistic epistemological position. *Nursing Philosophy, 8,* 102–113

Kinsella, E. A. (2012). Practitioner reflection and judgement as phronesis: A continuum of reflection and considerations for phronetic judgement. In E. A. Kinsella & A. Pitma (Eds.), *Phronesis as professional knowledge: Practical wisdom in the professions* (pp. 35–52). Rotterdam, Netherlands: Sense.

Lasica, J. D. (2013). Blogs and journalism need each other. *Nieman Reports.* Retrieved March 14, 2015 from http://niemanreports.org/articles/blogs-and-journalism-need-each-other/

Lind, R. A. (Ed.). (2012). *Produsing theory: The intersection of audiences and production in a digital world* (Digital Formations Series, Vol. 80). New York, NY: Peter Lang.

Lind, R. A. (2013). Laying a foundation for studying race, gender, class, and the media. In R. A. Lind (Ed), *Race/gender/class/media 3.0: Considering diversity across audiences, content, and producers* (3rd ed., pp. 1–12). Boston, MA: Pearson.

Malle, B. F., & Knobe, J. (1997). The folk concept of intentionality. *Journal of Experimental Social Psychology, 33,* 101–121.

Malle, B. F., Moses, L. J., & Baldwin, D. A. (2001). Introduction: The significance of intentionality. In B. F. Malle, L. J. Moses, & D. A. Baldwin (Eds.), *Intentions and intentionality: Foundations of social cognition* (pp 1–24). Cambridge, MA: The MIT Press.

McIntyre, R., & Smith, D. W. (1989). Theory of intentionality. In J. N. Mohanty & W. R. McKenna, (Eds.), *Husserl's phenomenology: A textbook* (pp. 147–179). Washington, DC: Center for Advanced Research in Phenomenology and University Press of America.

Ruykeyser, M. (1968). *The speed of darkness.* Retrieved February 28, 2015 from http://www.poetryfoundation.org/poem/245984

Saulauskas, M. P. (2000). The spell of *Homo Irretitus:* Amidst superstitions and dreams. *Information Research, 5*(4). Retrieved March 9, 2015 from http://informationr.net/ir/5-4/paper80.html

Schon, D. A. (1983). *The reflective practitioner: How professionals think in action.* New York, NY: Basic Books.

Schon, D. A. (1987). *Educating the reflective practitioner: Towards a new design for teaching and learning in the profession.* San Francisco, CA: Jossey-Bass.

Schon, D. A. (1992). The theory of inquiry: Dewey's legacy to education. *Curriculum Inquiry, 22*(2), 119–139.

Swartz, D. (2012). *Culture and power: The sociology of Pierre Bourdieu.* Chicago, IL: The University of Chicago Press.

Unger, R. M. (2001). *False necessity: Anti-necessitarian social theory in the service of radical democracy.* London, UK: Verso.

Zeichner, K. M., & Liston, D. P. (2013). *Reflective teaching: An introduction* (2nd ed.). New York, NY: Routledge.

The Interpretive Community Redux: The Once and Future Saga of a Media Studies Concept

Thomas R. Lindlof

New media and information technologies are often considered disruptive innovations because they roil existing industries and business models. Technological revolutions may also be responsible for disrupting certain theories, research programs, and seemingly settled issues of research practice. More than in most fields, scholars of media are at least moderately aware of the historical contingency of their theories—the notion that theories of mediated communication, although not exactly inhabiting the same cultural space as consumer fads and fashions, are certainly influenced by (and vulnerable to) the rise and fall of technological, economic, and political regimes, and thus should be regarded at least in part as intellectual artifacts of the epochs in which they are conceived.

Such is the case with audience studies. Since the turn of the millennium, a period coinciding roughly with the ascendancy of Web 2.0, audience researchers have witnessed the object of their inquiry radically transformed by the digital media revolution. Where once upon a time every mass medium controlled its own zone of content, effectively corralling the audiences seeking that content, digitalization has brought down those fences, probably forever. Where once media companies manipulated how and when audiences could gain access to their content, a seemingly limitless array of devices and apps equips people with the means to reel in content on their own timetables, creating new types of consumption events such as binge watching (Giuffre, 2013). Where once media narratives stayed put in their media of origin, stories are now re-circulated, re-mixed, and re-branded, in breathtakingly promiscuous fashion. Where once sharp distinctions existed between the production and consumption sides of the media culture equation, now anyone with modest technical skills—but perhaps immodest ambition—can be a star, auteur, or impresario of one's own channel. All while celebrities adopt the role of spectator and monitor from afar the conversations (about their careers, their body of work, their private lives, etc.) issued daily by fans encamped at digital firesides worldwide.

Audience researchers can be forgiven for reacting in a dazed and confused way to these head-snapping developments. After all, they have seen the term that identifies what they study—*audience*—cast in an increasingly anachronistic light, and with that realization comes gnawing doubts about the utility and relevance of their theoretical and methodological touchstones (Hermes, 2009; Jermyn & Holmes, 2006; Livingstone, 2004). It isn't that scholars lack ideas for dealing with the dilemmas they describe (e.g., Livingstone, 2013; Michelle, 2007). However, the quest for newer, better versions of audience studies in a vastly more complicated media ecosystem continues—a quest that includes a re-examination of familiar theories and concepts.

One area of audience studies—*interpretive community studies*—has survived the transition from mass to networked media. More than 30 years since its inception, the concept and the studies it has spawned are cited annually in hundreds of publications across all of the social sciences and humanities. The popularity of this idea stems at least in part from its protean quality of being able to assume different forms in fields from nursing to popular culture.

Its ongoing presence in media studies is also due to certain fundamentals of mediated communication that may never fade in importance. The shift from analog to digital, for example, hasn't altered the fact that interpretive activity transpires between people and mediated content. It can also be said, albeit with less certainty, that people still engage with media in predominately communal contexts. Indeed, "community" is a widely used trope for the social life flourishing in online sites. Nevertheless, it is worth asking how well suited the concept is for studying rapidly evolving (or mutating) forms of audience, and whether the version of interpretive community embraced by communication scholars in the era of mass media is the version we need for studying today's converged, on-demand media.

The first section of this chapter revisits the origins and intellectual underpinnings of the interpretive community idea and its adoption by media scholars. The second section reviews the recent proliferation of the interpretive community and related concepts, including issues of research practice that have limited its contributions. The final section addresses the prospects for studying the interpretive strategies of audiences (users, participants) in digital networks, particularly in contexts of transmedia storytelling and the produser culture.

The Interpretive Community Comes to Media Studies

The idea of the interpretive community came of age during the decade of the 1980s, a time when the communication discipline as a whole re-evaluated its positivist lineage and began exploring alternative perspectives. Its surfacing was part of a broader (and only partially successful) shift toward a cultural paradigm. Communication was relatively late to this interpretive turn, with the leading edge of social constructionism—a hydra-headed movement comprised of symbolic interactionism, social phenomenology, and ethnomethodology—having swept through sociology decades earlier. (The interpretive community was less firmly linked to the contemporaneous critical turn toward neo-Marxist, feminist, post-colonialist, and similar theories.)

In the United States, these influences first showed up in ethnographic case studies of television viewing (Lindlof, 2009). These studies, in one way or another, posed the question: How are media accommodated in the routines of everyday life? It wasn't long before a related question appeared: How do people make sense of media texts? Although cognitive psychology obviously had much to say about this question, for many it was best addressed by studying the discursive practices of sense-making. The impetus for this focus came from a host of British and Continental influences—semiology, cultural studies, and post-structuralism, in particular—but there was another major source of ideas that helped frame the future of audience research: literary theory.

Although some of the ground had been tilled earlier, by the likes of Charles Sanders Pierce, Georg Simmel, Max Weber, and Mikhail Bakhtin, it was the literary theorist Stanley Fish (1980) who coined the term "interpretive community" and fleshed out its core propositions. Like others in the reader-response movement, Fish challenged the authority of the text as the locus of meaning. Unlike most of them, he proposed a *social* solution to the question of how a text's meaning comes into being. As Fish described it, readers share certain strategies for interpreting and using text. Reading strategies, or coherent ways of approaching and making sense of a text, "exist prior to the act of reading and therefore determine the shape of what is read rather than, as is usually assumed, the other way around" (Fish, 1989, p. 115).

Crucially, Fish claimed that when a strategy is deployed, the reader actually writes the text by reading it. That is, the reader actively remakes the text in ways that are at once personal and social. In isolation, a text is just a set of inked marks (print) or light/sound emissions (electronic media); it comes to life as a meaningful object only when engaged by a reader who has learned (via acculturation into a community of like-minded readers) to apply the appropriate strategies of interpretation. Although shared strategies for reading text are the defining mark of such communities, there may occasionally be robust debates within a community over the practices and products of reading. Indeed, interpretive communities may quarrel not only over how to read (or write) a text but also about what *is* a text, how it should be valued and appraised, and so forth.

In Fish's formulation, an interpretive community is made up of a tight-knit, knowledgeable group of critics. The bridge to a more expansive view of interpretive community—a view that would encompass everyday readings of media and popular culture—was built initially through the work of Janice Radway. An American studies scholar, Radway was concerned with issues revolving around practical literacy: What do readers do with a text? Is reading "a singular, skilled process, which many readers only partially master, and some texts do not fully require"? (Radway, 1984a, p. 52). How do social relationships impinge on the activity of reading? How does the publishing industry grasp the interests of the book-buying public? And conversely, how does the public make sense of the genre categories, narrative conventions, and stylistic features favored by the industry?

Radway's seminal study, *Reading the Romance* (1984b), engaged these issues by taking Fish's ideas from the plane of the hypothetical to the world of actual readers, using interviews, text analysis, and other methods to explore the ways in which a group of women read and use romance novels in their lives. Radway discovered that women read romances for both the pleasures of feminine fulfillment and the relief it offered from the patriarchal relations and situations in their lives. She also claimed that women read the books as part of a community, both real and metaphorical. In its literal sense, the study revolved around a bookstore in the town of Smithton (a pseudonym) and focused on "a self-selected group of women united by their reliance on a single individual [the bookstore clerk Dorothy "Dot" Evans] who advises them about 'good' and 'bad' romances" (Radway, 1984a, p. 55). In its figurative usage, Radway inferred an active community of meanings and values from the women's verbalized readings of the romance text. Yet the relationship between the two levels of community was far from straightforward. Notably, Radway "was surprised to discover that very few of

[Dot's] customers knew each other . . . . I soon learned that the women rarely, if ever, discussed romances with more than one or two individuals" (1984b, p. 96). In other words, the similar readings of romance novels appeared to grow not so much out of the women's social networks (although Dot Evans did serve as a connector—a weak tie, in sociological parlance—between small groups of readers) but rather out of the imagined conversations they had with the authors. Radway elaborated on this view of community:

> The romance community, then, is not an actual group functioning at the local level. Rather, it is a huge, ill-defined network composed of readers on the one hand and authors on the other. Although it performs some of the same functions carried out by older neighborhood groups, this female community is mediated by the distances of modern mass publishing. Despite the distance, the Smithton women feel personally connected to their favorite authors because they are convinced that these writers know how to make them happy. (1984b, p. 97)

In addition to the achievement of revealing the women's understandings of this oft-denigrated genre, *Reading the Romance* signaled a new way of looking at media audiences. Instead of the media effects view of manifest content influencing the audience's cognitive schemas, an ideological view of the text's network of codes appealing to the reader's socio-cultural experiences, or a uses and gratifications view of satisfying the audience's conscious needs via media consumption, the interpretive community approach emphasized the role of social interaction. In other words, what audience members *do together* is create meaning from the raw materials of text: a bottom-up process of the invention, adoption, spread, and deployment of localized strategies for reading (viewing, listening to) media. In this formulation, the text isn't just a slate on which any meaning can be written. Every text is obviously put together in a certain manner, according to the purposes of its encoders (Hall, 1980). But once delivered to audiences, "the text becomes a site of contested interpretations with different audience communities producing different sense-making achievements" (Anderson & Meyer, 1988, p. 314). The approach, however, does not endorse a view of totally free, autonomous subjects who can read the text any way they want. As Fish wrote, the reader "proceeds from a collective decision as to what will count as literature, a decision that will be in force only so long as a community of readers or believers continues to abide by it" (1980, p. 11).

The years following the publication of Radway's book saw a sharp rise of interest in the concept among mass communication researchers. Among the first to begin theorizing about its role in media reception and usage was Klaus Bruhn Jensen (1987), who suggested that the notion of interpretive communities held the promise of "[serving] as a bridge between cultural and social science approaches to the media audience" (p. 29). According to Jensen, it could "differentiate the audience along lines of interests and usages" as well as account for the historical rise and fall of strategies of interpretation (p. 30)—ideas he continued to develop (Jensen, 1991, 1993).

Attention also turned to distinguishing the kinds of interpretive community that might exist. Lindlof (1988) proposed an analytic lens—consisting of genres of content, interpretation, and social action—by which to parse out, in prismatic fashion, the broad types of interpretive community operating in mass-mediated culture. With *genres of content,* people orient their reading strategies with respect to a certain category of text, becoming recognized as competent (or incompetent) readers in the eyes of others who use and interpret the same text (e.g., devotees of mockumentary films). This is the sense in which Fish wrote about literary critics and that Radway applied to romance readers. With *genres of interpretation,* on the other hand, people apply a strategy of interpretation across a wide variety of texts; this mode of sense-making, more often than not, springs from acculturation in a group's lived experience or identification with a belief system (e.g., evangelical Christians) and helps sustain the cohesion of the group itself. Finally, *genres of social action* locate reading strategy in a routine action or the performance of a social role (e.g., sharing certain pop culture information with one friend or a group of friends); somewhat like a pop-up store, this strategy may arise only when a specific occasion presents itself.

Clearly, media researchers were far more willing than Radway to think of this kind of community as an actual group—maybe not the classical notion of a geographically defined membership bound together by proximity, traditions, and obligations, but certainly a related conceptualization of social actors whose regular interactions forge similar understandings of text. This theorizing also suggests how similar readings of a mass-distributed text can coalesce locally, then spread widely, even rapidly, in overcoming the distance and anonymity of a nation-spanning audience—not unlike the manner in which worldwide transfers of scientific knowledge occur through what Paisley (1972) called "invisible colleges."

These ideas, however, were often left unfinished. So, perhaps predictably, this bold new theory, not yet tested or even explored empirically by many researchers circa early 1990s, was subjected to debate and criticism. Much of the critique revolved around the characterization of the media audience as a community. Schroder (1994) rejected the notion of an interpretive community based on socio-demographic categories or naturally occurring types of relationships (such as family), and advocated using the term only when media usage is the sole basis for group organizing (such as fandom). Machin and Carrithers (1996) advanced a rival concept, *community of improvisation,* to describe how people use media opportunistically in everyday situations. And, reversing his earlier support for the interpretive community concept, Jensen (2002) favored the term *interpretive repertoire* because it avoids the requirement of having to "belong" to "delimited communities" (p. 167).

Skepticism was also voiced about the concept's suitability for studying popular culture. Lichterman (1992) questioned whether "a concept originally applied to a literary establishment is as useful when applied to a loose network of readers" (p. 425). Elite literature, in this view, is deeply rooted in the specialized discourse and rigorous standards of readers who inhabit the life-world of professional criticism. Popular culture, on the other hand, is just a grazing ground for peripatetic readers—a culture too thin to support anything like the level of discussion about texts described by Fish.

The most insistent objection came from those who argued that interpretive approaches to media reception, especially when coupled with the notion of polysemy (Fiske, 1987), give nearly unlimited license to the agency of the audience. To these critics, it was a step too far toward radical postmodernism and ignored the ideological structures contained in media texts and the power of capitalist media to disseminate such texts and further exert their influence in society (Carragee, 1990). Often the critique went well beyond its initial target and became a blanket indictment of the populist or celebratory portrayals of audiences allegedly found in the qualitative research literature. Some audience scholars pushed back by citing misinterpretations of their work and disputing any necessary contradiction between powerful media and active audiences. The contentious back-and-forth of these debates reached its apogee in the mid-1990s, just about the time that the broader project of audience research began losing its luster in cultural studies (Clarke, 2000). By then, however, the term was firmly established in the media studies lexicon.

The Growth and Limitations of Interpretive Community Studies

Despite frequent use of the rubric "theory" for characterizing the interpretive community, including by myself, this usage isn't quite correct. Simply put, it is not theory, and never has been—certainly not in the Popperian sense of a set of propositions about the world that are capable of being refuted by empirical tests. Even with a more relaxed view of theory, current conceptualizations of interpretive community fail to tell us much about such basic matters as how to define its key components (e.g., strategy), where it can and cannot be applied (the boundary conditions of a theory), or the general mechanisms by which an interpretive community arises, changes, and sustains itself.

Despite being underdeveloped as a theory, the interpretive community has enjoyed a successful career as a *theoretical concept*—understood here as a tool for "disciplined imagination" (Alvesson & Karreman, 2007, p. 1266). In other words, the concept provides a vocabulary and a bundle of ideas for identifying and studying scenes in which the collective activity of reading texts is prominent. It has also proven useful for crafting hypotheses (Vannini, 2004) and designing studies of audience groups (Kretsedemas, 2010).

Another notable feature of the concept is its adaptability. It is a go-to concept in audience studies because it works well with a variety of *other explanations*—including social action, feminist and queer theories—and is adaptable to myriad cultural and social contexts. Among the types of audience community studied during the last 25 years are families (O'Guinn, Meyer, & McNeil, 1994), religious groups (Scott, 2003; Stout, 2004), fandom (Beck, 1995; Ellcessor, 2012; Hermes & Stello, 2000; Schaefer & Avery, 1993), political activists (Rauch, 2007), and ethnic and sexual minorities (Halse, 2012; Mikkonen, 2010; Mitra, 2010). The participants in some of these studies are aware (often acutely) of belonging to a group or a loose network in which media play an influential part. Others may not exhibit such awareness; in those cases, it is up to the researcher to establish a claim of community membership.

The contributions of this concept to knowledge of audiences are numerous. This chapter is not the place for a comprehensive review, but some findings are suggestive of its broader usage in the field. One stream of research concerns the *cultural influences on strategies of interpretation.* The objective in such studies is to explore ways in which the ideologies, value systems, lifestyles, or other kinds of cultural affiliation affect the interpretations of text. Scott (2003), for example, explored the influence of religious experience in the Church of Jesus Christ of Latter-day Saints on married couples' strategies of viewing popular television programs. Among other things, he found that some couples use a strategy of distancing themselves and their doctrinal values from the controversial topics featured in talk shows.

Other studies focus on *divergent community readings.* The text, in such instances, is a discursive site in which different readings are enacted by the participants. For example, in studying the narratives of visitors to the Field of Dreams baseball park outside Dyersville, Iowa (where *Field of Dreams* was filmed), Aden, Rahoi, and Beck (1995) developed three sets of paired themes that captured how people felt and thought when they "read the field" (p. 371): real/unreal, amusement/purpose, and community/isolation. Despite this evidence of divergent readings, the authors concluded that the visitors to the Field of Dreams "share the commonality of being drawn to the field by the movie and, in a deeper sense, by their identification with an idealized view of what the movie symbolizes" (p. 378).

Another area of study concerns *the usage of text in negotiating a reading community.* Here, research typically focuses on how the social roles and relationships of a community affect the activity of interpretation and vice versa. For example, a participant-observation study of three families over the course of several years (O'Guinn et al., 1994) revealed a number of changes in family VCR usage. Certain rituals for communal VCR readings—such as a family movie night—were co-created by family members, usually just after the device had come into the home. Family movie night, the authors noted, was "an evening . . . when the family consumed the benefits of the VCR's technological blessings together" (O'Guinn et al., p. 11). Over time, however-er, the family movie night eroded and the textual community it supported fell into disuse as family members felt "the powerful pull of other interpretive communities [outside the family]" that were a better fit for their individual interests, circumstances, and life experiences (O'Guinn et al., p. 20).

Interestingly, the interpretive concept has also been used by researchers to study media professionals who construct images of their work and collective identity through shared discourse (e.g., Berkowitz & TerKeurst, 1999; Bruggemann & Engesser, 2014; Zelizer, 1993). Indeed, it is a rare instance of a concept that has been found useful for studying both audiences and producers. This dual-purpose aspect of the concept is certainly beneficial in studying digital media produsers, as discussed in the next section.

As the interpretive community concept wended its way through media studies in the 1990s and 2000s, similar concepts surfaced and attracted interest in other fields; these include the epistemic community (Haas, 1992), the virtual community (Rheingold, 1993), the community of practice (Wenger & Snyder, 2000), and the brand community (Muniz & O'Guinn, 2001), which has had a profound impact in the field of marketing and consumer behavior. In many respects, the brand community bears a close kinship with the interpretive community of media studies; it describes the aficionados of a product (e.g., Macintosh computers), whose consumption rituals and traditions set them apart from mere purchasers or casual users.

Just as members of interpretive communities exercise agency of reading media texts, members of brand communities "represent a form of consumer agency" (Muniz & O'Guinn, 2001, p. 426). Brand communities engage the object of their affection with a fervent loyalty, a celebratory spirit, and a detailed practical knowledge of the product that is often generously shared with others in their social world. Membership in a brand community also brings a close affiliation with a commercial enterprise and often a long-term commitment to supporting the brand financially—aspects shared with some, but not all, interpretive media communities. The potential of the brand community and the related idea of tribal marketing (Cova & White, 2010; Hamilton & Hewer, 2010) for retrofitting the interpretive community concept for Web 2.0 are addressed later in the chapter.

Research of interpretive media communities has largely been conducted with qualitative methods: mostly interviews, and to a lesser degree, participant observation (Beck, 1995; O'Guinn et al., 1994; Tyler, 2010) and document analysis (Lindlof, 1996; Mitra, 2010). Much of the interview-based research of interpretive communities has adopted some version of *reception study* (or *reception analysis*), a text-reader model popularized by David Morley's (1980) influential study, *The "Nationwide" Audience.* Following Stuart Hall's encoding-decoding theory of the media communication process, in which the audience decoding stage invokes its own moments of meaningful determination quite separate from the encoding stage, reception study typically focuses on the audience's encounter with the signifying codes of a text. The text in such studies is carefully chosen to represent a theoretically significant category such as a genre or an ideological frame. In the typical protocol, the respondents are shown a sample of the text and asked a series of questions about it, such as what it meant to them, how they read it, how it compares to other texts, how it reflects (or doesn't reflect) their political views, cultural tastes, and so on. From analyzing the verbalized responses—that is, the evidence of respondent decoding of textual signifiers—the analyst infers the existence and functioning of an interpretive community.

Unfortunately, major limitations in our understanding of actual interpretive communities are largely due to its early capture by the reception study model. One problem stems from the way in which the activity of textual interpretation is represented. In reception study, the focal activity of audience is that of *reading from* (responding to) a text or genre. This conceptualization implicitly favors the text itself as the stable repository of meaning, thus shifting the analysis to questions of whether the audience's readings are denotative versus connotative or polyvalent versus polysemic (e.g., Condit,

1989; Michelle, 2007). Questions like these can serve useful purposes, such as distinguishing the various registers or modes of textual interpretation that may operate in encounters with a text. Yet the focus on response pulls us away from Fish's original insight that when readers engage with text, what they are really doing is *writing* their interpretations—not just figuratively but literally in the communicative forms of dialogues, monologues, debates, bodily expressions, symbolic artifacts, writings, artworks, and so on. Moreover, the reading-by-writing activity isn't summoned by textual signifiers. Rather, people actively construct a text, whether chosen intentionally or not, through the available strategies that matter to them. As such, texts are properly regarded as *resources* used by socially embedded readers to mobilize strategies and thus create and inscribe interpretations. The text-as-resource view doesn't rule out the role played in ideology or power as influential factors in the process; indeed, everyday acts of reading-by-writing can be complicit with, or take part in struggles against, the larger forces shaping the industrial production and distribution of mediated content.

Another problem with the reception study model is its limited grasp of how actual communities operate. As Anderson remarked many years ago, "The successful analysis of the [interpretive community] must enlarge its scope well beyond the point of contact with mediated content" (1996, p. 75). This need remains unchanged. Few, if any, interpretive communities subsist on a single text. They are not monocultures. Rather, they thrive on a wealth of related textual (and intertextual) materials and cultural performances, and respond to events and pressures from outside their borders. Enlarging the scope of analysis beyond the point of contact means, for example, studying people who have never laid eyes on a particular text, yet who still participate discursively in the community. It also means considering the multiple ways in which people engage with a text, with all of the technological means available to them. In addition, the successful analysis needs to consider the arc of a community's strategies and social relationships over time, and thereby explore the dynamic, often unpredictable, shape-shifting of meaning in situ.

The problems cited here, inherited largely from the unreflective use of the reception study model, have held back a deeper understanding of the interpretive community phenomenon in everyday life. Yet these problems must be overcome—and the reception model either improved upon or set aside—if the concept is to be productively applied beyond its old media origins.

Toward Produser Communities

The foregoing discussion suggests that the interpretive community is both an appropriate and an inadequate concept for studying digital culture. The concept is appropriate because it presumes that people *write* their readings in dynamic social interaction. In fact, more than any other concept in audience research, the interpretive community concept is well suited for studying *produsage* (Bruns, 2008)—the idea that audiences not only use mediated content, but are also motivated and empowered to produce content in a variety of media platforms, formats, and discourses.

However, the interpretive community concept is also inadequate because of its traditional focus on an isolated text that stays the same for all audience members. This focus is dramatically at odds with the nature of transmedia storytelling and the mobility of reading/writing strategies within and across platforms. For the concept to adapt to the new realities of participatory culture and hence become truly useful for media scholars, it must propose new ways of asking and addressing questions about the meaning-making activity of audiences (users, participants).

To begin with, the interpretive community concept must demonstrate its capacity for studying and delivering insights about the forums that surround, interpenetrate, and comment on industrially produced media texts. To put this as a question, how do digital media forums function as communities of text produsers? The task is admittedly complex and fraught with methodological challenges. Addressing it depends on the specific forums, texts, and strategies singled out for study. Two recent studies provide excellent—and contrasting—starting points. The first is Mitra's (2010) study of several queer Indian blogs that sprang into high gear following Gay Pride protest marches held simultaneously in three Indian cities in June 2008. In the immediate wake of these marches, "queer Indian bloggers . . . solicited comments, setting into motion a veritable cycle of interpretation. Many of the bloggers/readers/commenters were present at the marches, and narrated how the media coverage corresponded with their lived experience . . . [thus] acting as both a media audience as well as producers" (p. 165). The texts under study were quite diverse and materialized in both offline and online settings: the mainstream news accounts of the marches, the commentary circulating among and between bloggers and their audiences, and the blogs' linkages to other sites (e.g., YouTube clips). The complexity and dynamism of community-building strategies were equally striking. The author documented, among other things, the hyperlinking with other queer groups to spread awareness and invite participation beyond the blogs' regular audience; the bloggers' reportage of their own participation in the day's events, which enhanced their credibility and community standing; the bloggers' efforts to denaturalize the news media's frame for the readers, thus assisting in the co-creation of a critical sensibility; and vigorous dialogues among the bloggers and their readers about the stereotypes employed in the news coverage—a core concern of Indian gay culture. Interestingly, the Gay Pride marches were read (written) very differently by other segments of the blogosphere—not just as protest but also as a spectacle, a celebration, or an occasion for enlightenment. "[N]one of the blogs studied," observed Mitra, "used either/or of the multiple strategies, and usually employed a *combination* of them" (emphasis in original; p. 175). Ultimately, the study shows how a blog can function as "a tool for both community-creation and self-writing" (p. 167), particularly in highly charged moments when a stigmatized minority is the subject of widespread scrutiny.

The contrasting example is Wood and Baughman's (2012) investigation of *Glee* fan practices on Twitter. The authors wanted to understand "how consumer fans of *Glee* who maintain Twitter 'character' accounts actively participate in transmedia storytelling through social interactions with other fictional accounts belonging to the show's 'characters' as well as lower level fans" (p. 330). The text in this case is an ongoing TV series (instead of news coverage of a one-off public event). The forum is the social-networking service in which account holders tweet to followers in real time (rather than asynchronous blog posts). The fans' strategies are governed by rules and conventions they have jointly established (rather than the more individualistic, free-wheeling spirit of the Indian blog posters and commenters).

These fans are clearly a subset of the broader community of regular *Glee* viewers. They are comfortable with and skilled at the art of engaging in role-play during real-time viewing of *Glee*. Role-playing a *Glee* character "reproduce[s] an older form of television viewing that allows marketers to recapture them as an audience" (Wood & Baughman, 2012, p. 338), while also forcing the fans to "follow the directed script of the show as their only source for information on the personalities of [the] characters" (p. 339). However, the tweets flashing back and forth among the fans do not just follow the show's storylines or mimic how the characters behave from one moment to the next. Nor are the tweets just idle chats, disconnected from what is occurring in the program (although this did happen occasionally). Rather, the authors mainly found a process they called "narrative augmentation." In effect, the fans creatively *rewrite* the show's text at the same they read it. They augment the *Glee* narrative in a variety of ways: actively monitoring the show for plot points, referencing events their characters were involved in, "[adding] commentary through character voice" (p. 338), reacting to the role-play of another account holder, and more. Ultimately, the work of managing a character involved the fans in a sort of interpretative steering process that the authors called "identity control"—a process by which the fans highlight certain traits (language, topics, etc.) to maintain the consistency and integrity of the character. An effective performance of identity control appears to mark the account holder as a knowledgeable participant in this community.

What can be learned about user forums as an interpretive community from the two studies? First, the traditional audience role continues to exist. The people in these forums read industrially produced texts expecting to find value and coherence in the experience. However, they also use the online forums to reply to others who read the selfsame text, and to augment the source text and recast its meanings. Arguably, these strategies for writing interpretations ramp up the felt sense of community among the participants. In fact, the experience of circulating information and forging connections with like-minded others may be a major reason for consuming text.

The digital-culture iteration of the interpretive community concept must also come to grips with transmedia storytelling—a need hinted at in the cases examined earlier. The economy of mass media, based on building silos of copyrighted works and exploiting the audience labor of viewing or listening to them, has not vanished, but by almost any reckoning, its days of dominion are, if not numbered, in danger of yielding to the social advantages inherent in Web 2.0. Indeed, most media companies have already pivoted away from the model of single-screen, pushed content. In its place have arisen business models that give audiences multiple opportunities to pull the text of their choice, or several versions of a text, to virtually any device. If every device has its own affordances, it is eminently sensible to craft a narrative product into distinctive, albeit interrelated, versions: *transmedia.*

Transmedia story production, in turn, thrives in symbiotic relationship to two other concepts central to digital media: the *gift economy,* an ethos of sharing industrially-produced and user-created symbolic goods, in which status accrues to those who are generous; and the *collective intelligence* of fans and other dispersed users in creating symbolic goods (Jenkins, Ford, & Green, 2013; Levy, 2013). In the economic (and moral) paradigm encompassing all three of these concepts, the value of any symbolic good is heightened by its shareability and the ability of those who share it to personalize it in some way. Similarly, as Mittell (2013) noted, a key attribute of transmedia, adapted from wiki websites, is that of *emergence:* "a bottom-up phenomenon, coming together through the collection of small practices" (p.

37). Such practices range from the truly small (e.g., brief comments, annotations) to the not-small-at-all (e.g., full-blown remixes). In social-networking sites, for example, posted content is virtually a failure if it isn't passed along or followed, or if it goes out without attracting some degree of co-authorship from other people's comments, likes, tags, retweets, and so forth. Ultimately, these practices of adding, editing, and curating material may be kept alive for as long as people collectively perceive some sort of value to be had from engaging with it. The long tail of the digital economy may be matched by a cultural long tail: a vision of incredibly numerous interpretive communities receding into the distance, each of them clinging to an aging text, updating the interpretive strategies applied to it, and sharing their passions online.

One major implication of this world of proliferating versions of texts is that a study based on the audience's response to a single representation, as is the habit with reception studies, is grossly inadequate. But that is the easiest, most obvious issue to dispose of, and once it is off the table, a host of more important—and much more vexing—issues flood the zone: Where is the community located if it isn't attached to a single text (genre)? If each version of a transmedia text engenders its own group of readers, how do their interpretive strategies differ or overlap? How, in fact, do we draw the lines around the core strategies for interpreting a transmedia story? When a commercially successful text begins to scale up its transmedia power and reach, does the original interpretive community scale up as well, or does rapid growth spin off different (perhaps innovative, heretical, or reactionary) communities? If the media industries assign hierarchies of economic value to the versions they produce, as they often do, do interpretive communities create their own hierarchies, perhaps based on very different value propositions concerning transmedia texts? And so on.

Although the full weight of these research agendas can stagger the imagination (or fire up the imagination, for the brave among us who seek a genuine challenge), one can at least consider the first question and potentially get somewhere with it: Where is the community located? Several interesting ideas have surfaced recently. In their essay on the limits of fan authorship, Stein and Busse (2009) identified interpretive communities as a major limiting factor on fandom, stating that "fan communities constitute discursive contexts that join the official source text as intertextual referent" (p. 196). What they called the "fannish intertext"—the works of fan authorship created around a character, trope, or story line—often "shapes a given fan community's expectations" (p. 197). In this view, the fans themselves decide what part of a text deserves to be the focus of discourse and thus the basis for organizing a community:

Fannish interpretive communities define themselves around shared readings of a character, a pairing, or a particular aspect of a fictional universe. Communities may form around central interpretive moments such as the celebration or rejection of a central plot point or a particularly aggressive reading of a controversial source text event . . . . Preferences for particular romantic pairings also clearly delineate interpretive communities; indeed, many fans identify themselves primarily as fans of one or another pairing. As such, they agree on the centrality of particular events, characteristics, and interpretations that support their favored romantic pairing. (p. 197)

As this passage suggests, the full text isn't always the basis for organizing community. Through Internet-enabled devices and spaces, fans can find other fans who share a specific, even esoteric, interest in a compelling part of a text. Although Stein and Busse (2009) focused on creators of fan fiction, art, music, and so forth, their argument can be extended to less intensive (or less artistically inclined) forums that pop up in the general population. Analysis of these forums, according to Hamilton and Hewer (2010), can "[reveal] a sense of an emergent computer-mediated emotional community, where we witness the shared affiliation and social ties that are expressed through talk around [a text]" (p. 279).

Another way of thinking about the location of interpretive community in digital media is to consider the social linkages—or connectors—*between* texts. That is, rather than focus just on the industrially produced text, or only the texts produced by users (fans), one can focus on the dynamic interplay between types of texts, or the paths by which users move into, through, and out of the discursive space, with the attendant possibilities for growth, mutation, and/or transformation of interpretive strategies. For example, Ellcessor (2012) studied the "star text of connection" in the world of celebrity, arguing that "websites and their users are best studied through their connections to other sites, services, individuals, and communities" (p. 49). The celebrities themselves establish connections with fan communities via "free blogging software, low-cost image and video tools, video sharing sites, and social networking sites," as the means of "shaping a celebrity persona" (p. 59). Similarly, marketing-oriented studies of word-of-mouth communications offer intriguing glimpses of how inter-community affiliation and conflict may occur in social networks of popular culture (Aral, 2013; Kozinets, de Valck, Wojnicki, & Wilner, 2010; Thomas, Price, & Schau, 2013).

Finally, it is worth being reminded that many interpretive communities attach themselves to brands rather than to a particular text. Obviously, private sector companies recognize that they have a great deal at stake in growing and sustaining the interest, if not loyalty, of brand communities. In attempting to create relationships with members of these communities and gain access to what they have to say about brands (all of it fodder for sentiment analysis and other types of data mining/analytics), companies often try to go native—mingling in this environment, offering free content, setting up shop inside people's mobile devices and social media profiles, and otherwise using any available tactic to promote their brands in these social spaces. Media studies would certainly profit from deeper investigation of the commercial underpinnings of interpretive communities.

The Way Forward

In retrospect, it may seem a happy accident that the idea of the interpretive community appeared in the 1980s when the field of mass communication needed it the most. However, it is also the case that new theories or concepts can cross-pollinate several fields at more or less the same time, especially when disparate groups of scholars view them as important for solving similar problems. With a turn toward a hermeneutical science (Taylor, 1971) underway in all of the social sciences, media audience scholars in the 1980s were primed for a new approach to a growing problem: specifically, the problem of how meanings of mass-mediated content are socially created. The interpretive community concept helped clarify this problem and gave audience researchers a rich source of ideas for conducting empirical studies. That the concept remained just a concept, rather than developing into a theory, is an interesting part of the story. However, it does not negate the contributions of the concept, nor is it a reason for theoretical explanations of interpretive communities to stay undeveloped in the future.

As discussed earlier, technological changes often usher in (or force) new ways of theorizing media phenomena. For example, there were television-defined communities before the video cassette recorder came along (Hutchison, 2012), but the timely introduction of the interpretive community concept provided audience researchers with a vocabulary and conceptual tools for studying how people preserve and share video texts and organize social rituals around VCR technology. Similarly, the advent of Web 2.0 brought a new set of challenges for understanding media usage. In one sense, digital networks and multimedia devices allow people to share interpretive strategies in much the same way as before—only faster, more efficiently, and with fewer physical obstacles. But these features, coupled with the ability to write interpretations in a variety of multimedia formats—in effect, adopting the produser identity—are part of a profoundly transformative stage in the progression of the audience in mediated culture.

The interpretive community is a mature concept, yet it remains a promising tool of disciplined imagination for studying this transformative stage. The concept has helped us understand how user forums organize themselves as self-reflexive communities, and has shown that reading in the digital media landscape is a multiplex skill set involving knowledge of many codes, contexts, and competencies. It also enables us to understand, at this early stage of research, how people form relationships to a source text through their communal activities. There is still a long way to go. Methodological strategies for tracking and understanding media usage in online (and combined online-offline) contexts are still being developed (Picone, 2013; Postill & Pink, 2012). And although it is always tempting to regard the leading front of technological advancement as the zeitgeist, the more accurate picture of everyday media usage is one of bricolage. That is, most of us piece together all sorts of media of varying historical provenance—such as smartphones, AM/FM radio, MP3 players, tablets, traditional television, IMAX movies, newspapers or magazines—to achieve immediate goals or to connect ourselves to domains of mediated experience. The challenge of studying these issues through revision of the interpretive community concept is difficult, but the payoff in greater understanding may be worth the effort.

References

Aden, R. C., Rahoi, R. L., & Beck, C. S. (1995). "Dreams are born on places like this": The process of interpretive community formation at the Field of Dreams site. *Communication Quarterly, 43,* 368–380.

Alvesson, M., & Karreman, D. (2007). Constructing mystery: Empirical matters in theory development. *Academy of Management Review, 32,* 1265–1281.

Anderson, J. A. (1996). The pragmatics of audience in research and theory. In J. Hay, L. Grossberg, & E. Wartella (Eds.), *The audience and its landscape* (pp. 75-93). Boulder, CO: Westview Press.

Anderson, J. A., & Meyer, T. P. (1988). *Mediated communication: A social action perspective.* Newbury Park, CA: Sage.

Aral, S. (2013). What would Ashton do—And does it matter? *Harvard Business Review, 91*(5), 25–27.

Beck, C. S. (1995). You make the call: The co-creation of media text through interaction in an interpretive community of "Giants' fans." *The Electronic Journal of Communication, 5*(1). Retrieved January 5, 2015 from http://www.cios.org/EJCPUBLIC/005/1/00515.HTML

Berkowitz, D., & TerKeurst, J. V. (1999). Community as interpretive community: Rethinking the journalist-source relationship. *Journal of Communication, 49*, 125–136.

Bruggemann, M., & Engesser, S. (2014). Between consensus and denial: Climate journalists as interpretive community. *Science Communication, 36*, 399–427.

Bruns, A. (2008). *Blogs, Wikipedia, Second Life, and beyond: From production to produsage.* New York, NY: Peter Lang.

Carragee, K. M. (1990). Interpretive media study and interpretive social science. *Critical Studies in Mass Communication, 7*, 81–96.

Clarke, D. (2000). The active pursuit of active viewers: Directions in audience research. *Canadian Journal of Communication, 25*(1), 39–59. Retrieved December 16, 2014, from http://www.cjc-online.ca/index.php/journal/article/view/1138/1057

Condit, C. M. (1989). The rhetorical limits of polysemy. *Critical Studies in Mass Communication, 6*, 103–122.

Cova, B., & White, T. (2010). Counter-brand and alter-brand communities: The impact of Web 2.0 on tribal marketing practices. *Journal of Marketing Management, 26*, 256–270.

Ellcessor, E. (2012). Tweeting @feliciaday: Online social media, convergence, and subcultural stardom. *Cinema Journal, 51*(2), 46–66.

Fish, S. (1980). *Is there a text in this class?* Cambridge, MA: Harvard University Press.

Fish, S. (1989). Interpreting the variorum. In R. C. Davis & R. Schlifer (Eds.), *Contemporary literary criticism* (pp. 101–117). New York, NY: Longman.

Fiske, J. (1987). *Television culture.* London, UK: Methuen.

Giuffre, L. (2013). The development of binge watching. *Metro, 178*, 101–102.

Haas, P. (1992). Introduction: Epistemic communities and international policy coordination. *International Organization, 46*(1), 1–36.

Hall, S. (1980). Encoding/decoding. In S. Hall, D. Hobson, A. Lowe, & P. Willis (Eds.), *Culture, media, language* (pp. 128–138). London, UK: Hutchinson.

Halse, R. (2012). Negotiating boundaries between us and them: Ethnic Norwegians and Norwegian Muslims speak out about the "next door neighbour terrorist" in 24. *Nordicom Review, 33*(1), 37–52.

Hamilton, K., & Hewer, P. (2010). Tribal mattering spaces: Social-networking sites, celebrity affiliations, and tribal innovations. *Journal of Marketing Management, 26*, 271–289.

Hermes, J. (2009). Audience studies 2.0. On the theory, politics and method of qualitative audience research. *Interactions: Studies in Communication and Culture, 1*(1), 111–127.

Hermes, J., & Stello, C. (2000). Cultural citizenship and crime fiction: Politics and the interpretive community. *European Journal of Cultural Studies, 3*(2), 215–232.

Hutchison, P. J. (2012). Magic windows and the serious life: Rituals and community in early American local television. *Journal of Broadcasting & Electronic Media, 56*, 21–37.

Jenkins, H., Ford, S., & Green, J. (2013). *Spreadable media: Creating value and meaning in a networked culture.* New York: New York University Press.

Jensen, K. B. (1987). Qualitative audience research: Toward an integrative approach to reception. *Critical Studies in Mass Communication, 4*, 21–36.

Jensen, K. B. (1991). When is meaning? Communication theory, pragmatism, and mass media reception. In J. A. Anderson (Ed.), *Communication yearbook 14* (pp. 3–32). Newbury Park, CA: Sage.

Jensen, K. B. (1993). The past in the future: Problems and potentials of historical reception studies. *Journal of Communication, 43*(4), 20–28.

Jensen, K. B. (2002). Media reception: Qualitative traditions. In K. B. Jensen (Ed.), *A handbook of media and communication research: Qualitative and quantitative methodologies* (pp. 156–170). London, UK: Routledge.

Jermyn, D., & Holmes, S. (2006). The audience is dead: Long live the audience! Interactivity, "telephilia," and the contemporary television audience. *Critical Studies in Television, 1,* 49–57.

Kozinets, R. V., de Valck, K., Wojnicki, A. C., & Wilner, S. J. S. (2010). Networked narratives: Understanding word-of-mouth marketing in online communities. *Journal of Marketing, 74*(2), 71–89.

Kretsedemas, P. (2010). "But she's not black!" Viewer interpretations of "angry black women" on prime time TV. *Journal of African American Studies, 14,* 149–170.

Levy, P. (2013). The creative conversation of collective intelligence (Trans., P. Aronoff & H. Scott). In A. Delwiche & J. J. Henderson (Eds.), *The participatory cultures handbook* (pp. 99-108). New York, NY: Routledge.

Lichterman, P. (1992). Self-help reading as a thin culture. *Media, Culture and Society, 14,* 421–447.

Lindlof, T. R. (1988). Media audiences as interpretive communities. In J. A. Anderson (Ed.), *Communication Yearbook 11* (pp. 87–107). Newbury Park, CA: Sage.

Lindlof, T. R. (1996). The passionate audience: Community inscriptions of *The Last Temptation of Christ.* In D. A. Stout & J. Buddenbaum (Eds.), *Mass media and religion: Audiences and adaptations* (pp. 148–168). Thousand Oaks, CA: Sage.

Lindlof, T. R. (2009). Qualitative methods. In M. B. Oliver & R. Nabi (Eds.), *The Sage handbook of media effects* (pp. 53–66). Thousand Oaks, CA: Sage.

Livingstone, S. (2004). The challenge of changing audiences: Or, what is the audience researcher to do in the age of the Internet? *European Journal of Communication, 19*(1), 75–86.

Livingstone, S. (2013). The participation paradigm in audience research. *The Communication Review, 16,* 21–30.

Machin, D., & Carrithers, M. (1996). From "interpretative communities" to "communities of improvisation." *Media, Culture & Society, 18,* 343–352.

Michelle, C. (2007). Modes of reception: A consolidated analytical framework. *Communication Review, 10,* 181–222.

Mikkonen, I. (2010). Negotiating subcultural authenticity through interpretation of mainstream advertising. *International Journal of Advertising, 29,* 303–326.

Mitra, R. (2010). Resisting the spectacle of pride: Queer Indian bloggers as interpretive communities. *Journal of Broadcasting & Electronic Media, 54,* 163–178.

Mittell, J. (2013). Wikis and participatory fandom. In A. Delwiche & J. J. Henderson (Eds.), *The participatory cultures handbook* (pp. 35–42). New York, NY: Routledge.

Morley, D. (1980). *The "Nationwide" audience.* London, UK: British Film Institute.

Muniz, Jr., A. M., & O'Guinn, T. C. (2001). Brand community. *Journal of Consumer Research, 27,* 412–432.

O'Guinn, T. C., Meyer, T. P., & McNeil, M. (1994). *The family VCR: Ordinary family life with a common textual product.* Unpublished manuscript.

Paisley, W. (1972). The role of invisible colleges in scientific information transfer. *Educational Researcher, 1*(4), 5–8, 19.

Picone, I. (2013). Situating liquid media use: Challenges for media ethnography. *Westminster Papers, 9*(3), 49–70.

Postill, J., & Pink, S. (2012). Social media ethnography: The digital researcher in a messy web. *Media International Australia, 145,* 123–134.

Radway, J. (1984a). Interpretive communities and variable literacies: The functions of romance reading. *Dedalus, 113*(3), 49–71.

Radway, J. (1984b). *Reading the romance.* Chapel Hill: University of North Carolina Press.

Rauch, J. (2007). Activists as interpretive communities: Rituals of consumption and interaction in an alternative media audience. *Media, Culture & Society, 29,* 994–1013.

Rheingold, H. (1993). *The virtual community: Homesteading on the electronic frontier.* Cambridge, MA: The MIT Press.

Schaefer, R. J., & Avery, R. K. (1993). Audience conceptualizations of *Late Night with David Letterman. Journal of Broadcasting & Electronic Media, 37,* 253–273.

Schroder, K. C. (1994). Audience semiotics, interpretive communities and the "ethnographic turn" in media research. *Media, Culture & Society, 16,* 337–347.

Scott, D. W. (2003). Mormon "family values" versus television: An analysis of the discourse of Mormon couples regarding television and popular media culture. *Critical Studies in Media Communication, 20,* 317–333.

Stein, L., & Busse, K. (2009). Limit play: Fan authorship between source text, intertext, and context. *Popular Communication, 7,* 192–207.

Stout, D. (2004). Secularization and the religious audience: A study of Mormons and Las Vegas media. *Mass Communication and Society, 7*(1), 61–75.

Taylor, C. (1971). Interpretation and the sciences of man. *Review of Metaphysics, 25,* 3–34, 45–51.

Thomas, T. C., Price, L. L., & Schau, H. J. (2013). When differences unite: Resource dependence in heterogeneous consumption communities. *Journal of Consumer Research, 39,* 1010–1033.

Tyler, J. (2010). Media clubs: Social class and the shared interpretations of media texts. *Southern Communication Journal, 75,* 392–412.

Vannini, P. (2004). The meanings of a star: Interpreting music fans' reviews. *Symbolic Interaction, 27*(1), 47–69.

Wenger, E. C., & Snyder, W. M. (2000). Communities of practice: The organizational frontier. *Harvard Business Review, 78*(1), 139–145.

Wood, M. M., & Baughman, L. (2012). *Glee* fandom and Twitter: Something new, or more of the same? *Communication Studies, 63,* 328–344.

Zelizer, B. (1993). Journalists as interpretive communities. *Critical Studies in Mass Communication, 10,* 219–237.

CHAPTER THREE

Duality Squared: On Structuration of Internet Governance

Dmitry Epstein

On Wednesday, January 18, 2012, the Internet went dark, in many cases literally. The English Wikipedia, Reddit, Google, Flickr, and others—together over 115,000 websites—presented some kind of banner or landing page to protest two laws proposed in the U.S. Congress. The Stop Online Piracy Act (SOPA) and Preventing Real Online Threats to Economic Creativity and Theft of Intellectual Property Act (PIPA) were aimed at curbing copyright and intellectual property violations on websites hosted outside of the United States by focusing enforcement at the level of the Domain Name System (DNS). What made those protests unique was the very public way in which the technical community confronted policy makers and the public support they were able to garner. This is a vivid illustration of how structure and agency play out in Internet governance through technology, culture, and policy.

The SOPA/PIPA protests illustrate fundamental tension and inherent interdependency between the East Coast code and the West Coast code (Lessig, 2006). Following a U.S. geographic metaphor, the former refers to laws and regulations and the latter to computer and web programs and technical standards. This time, the East Coast versus West Coast tension came to a boiling point, and Silicon Valley openly and publically engaged the Hill. As John Battelle, cofounder of *Wired,* put it: "We can't afford to not engage with Washington anymore . . . . Silicon Valley is waking up to the fact that we have to be part of the process in Washington—for too long we've treated 'Government' as damage, and we've routed around it" (2012). But the West Coast engaging with the East Coast is only part of the story. It is still only a tale of elites trying to mold structures that preserve their power and further serve their interests (Genieys & Smyrl, 2008).

To complete the picture, one has to account for the cultural norms that evolved around the use of the Internet and online civic engagement, as well as for the continuously evolving affordances of the Internet itself and the numerous applications on its edges (Bridy, 2012). For example, on the day of the blackout, more than a million messages were sent to the members of Congress via an online tool offered by the Electronic Frontier Foundation, 2.4 million tweets about SOPA were posted on Twitter, and over 4.5 million people signed a petition started by Google (Netburn, 2012; Samuels & McSherry, 2012). Whether it was that public outcry that stalled the SOPA/PIPA legislation or the resources provided by the large technology companies, people speaking out on technology regulation at such scale was at that point unprecedented.

The SOPA/PIPA showdown is a relatively rare yet vivid example of how various actors and structural arrangements play into the constitution of information society. This chapter puts forward a proposal for a duality squared model—a structuration-theory-based framework to analyze the interaction among information technology artifacts, their users, designers, and policy makers regulating information governance, as well as the policy artifacts (regulations and regulatory institutional settings) they create. Conceptually, this proposal is motivated by my interest in the inherent tension between individual agency and micro-behaviors of individuals and the systemic and structural properties of the environments in which infor-mation technologies are created, regulated, and used. Practically, this work is fueled by the ongoing discussions about Internet governance and the growing body of literature on this topic (DeNardis, 2010). Technologies and policies governing how information can be created, used, shared, remixed, abused, and so on, make a particularly interesting case for the analysis of this tension both because of their ubiquitous presence in contemporary society and because of their fundamental importance for the notion of power in social analysis (Braman, 2009a).

Internet, Governance, and Society

The politics of the Internet are enacted through the numerous creative and disruptive ways this technology has been and is being used. Some scholars argue that the politics of the Internet are inherent in its design. Laura DeNardis (2009), for example, noted how the engineering of the network embodied choices about civil liberties such as privacy and freedom of speech: "Internet architecture and virtual resources cannot be understood only through the lens of technical efficiency, scarcity, or economic competition but as an embodiment of human values with social and cultural effects" (p. 96). Others focus on the enabling aspects of a network, which, based on libertarian ideas, transcended traditional boundaries of state control of media and communication channels. Mueller (2010) argued that the Internet "changes the polity" by altering "the cost and capabilities of group action" and enabling "new forms of collaboration, discourse, and organization" (p. 5), which in turn allows new forms of transnational governance. The Internet allowed unprecedented political mobilization by realigning the technical basis of what Braman (2009a) labeled "informational power"—the informational origins "of the materials, social structures, and symbols that are the stuff of power in its other forms" (p. 26). The ability to innovate, whether politically, commercially, or socially, on the edges of the network shifted the balance of political power between the state and the individual.

Governing the Internet imposes politics on this complex sociotechnical system. Internet governance plays out as politics of control when it comes to management and distribution of domain names and IP addresses, and stirs "questions about how access to resources and power over these resources are distributed or should be distributed among institutions, nation-states, cultures, regions, and among entities with a vested economic interest in the possession or control of these resources" (DeNardis, 2009, p. 16; see also Galloway, 2006). Internet governance also plays out as cultural politics in a debate about what values and core principles should be preserved as the network changes. Influencing the technical infrastructure of the Internet means influencing the civil liberties that are enacted through this technology (Braman, 2011).

Today, Internet governance is referred to not only as governance of the technical infrastructure but also as control of online behaviors or the very enactment of the liberties it affords (Mueller, 2010). As such, Internet governance also plays out as global politics of domination. Nation states and regional and international alliances are competing for the establishment of legal frameworks and public policy practices that preserve the national interests and value systems of the parties involved. The long history of cultural, political, and economic tensions among nation states is reinterpreted within the Internet governance debate, thus making it also a debate about values of democratic participation, economic freedoms, and cultural hegemony (Hart, 2011).

Throughout this chapter, I use a rather broad but well-defined meaning of "governance" as "decision-making with constitutive [structural] effect whether it takes place within the public or private sectors, and formally or informally" (Braman, 2009a, p. 3). This is to differentiate governance from the narrow meaning of government and the conceptually different idea of governmentality (Braman, 2009a, p. 3). The processes of governance—such as legislation, corporate policy making, or articulation of community norms—are constitutional social forces. They organize existing social categories and relationships, and they define new social categories within the context of already existing systems of rules and institutions. Thus governance is a continuous and conscious act of social construction or, expanding Fischer and Forester's (1993) definition of policy making, ". . . a constant discursive struggle over the criteria of social classification, the boundaries of problem categories, the intersubjective interpretation of common experiences, the conceptual framing of problems, and the definition of ideas that guide the ways people create the shared meanings which motivate them to act" (pp. 1–2). Law and policy both trigger and react to social change, so "with a longer and wider view it is possible to see a specific law developing out of cultural practice, becoming a form of discourse, and ultimately being translated into technology" (Braman, 2009a, p. 3).

Information policy, or more broadly, governance of information, adds a layer of complexity to the dualistic relationship between policy and society. First, this complexity stems from the omnipresence of information—it is both a constitutive social force and a fundamental component of governance. Capturing the duality of agency and structure within this dynamic relationship is one of the main challenges in theorizing Internet or information governance. Second, the dualistic relationship between information policy and society is mediated through technology use. Formal Internet-related policy making, particularly that conducted by governments, often lags behind not only corporate decision making regarding creation and management of information tools and resources but also the users' ever-evolving patterns of use. Thus, unpacking the social constructive forces surrounding technology creation, adoption, and use are pivotal to understanding the Internet, information policy, and governance.

There is, however, a disconnect between attempts to conceptualize and critique Internet governance processes and institutions and attempts to conceptualize technology adoption. This disconnect is particularly evident when one is trying to focus on the duality of agency and structure. The literature in Internet governance draws mainly on theories of institutional economics and international relations (DeNardis, 2010; Mathiason, 2009;

Mueller, 2010; Singh, 2009), with only a few drawing on science and technology studies (DeNardis, 2009; Flyverbom, 2011; Mueller, Kuehn, & Santoso, 2012). Across the board, the primary focus of Internet governance literature is on *institutions* as political actors or as constraining factors in decision-making processes. Development of technology is typically treated as either exogenous or constrained by institutional forces. Moreover, while individual actors and their actions are acknowledged, the accounts are historical in nature and there is no explicit discussion of agency.

Conceptualizations of technology adoption and use present more nuanced considerations of the duality of agency and structure. Most prominently, Orlikowski (1992, 2000) and then DeSanctis and Poole (1994) successfully adapted the theory of structuration (Giddens, 1984) to explain information technology adoption and change in organizational settings. The structurational model of technology views technology as both a product and a medium of human action, both occurring within institutional context and having consequences for institutional properties. More specifically, while human agents and technological artifacts are viewed as mutually influential, technology is conceptualized as impacting institutional properties of an organization, while those properties impact human agents (Orlikowski, 1992). This is a powerful model that steers away from the exogenous treatment of technology and views it instead as a consequence of human activity.

Missing from the structurational view of information technology is a clear articulation of policy as a structural element that is both a product and a medium of human action, with clear dependency and influence over social structures. We need a theoretical model that brings together the *structural aspects* of policy, technology, and human behavior vis-à-vis information and information technology, with *individual agency* in shaping these policies, technology, and behavioral norms. In this chapter, I attempt to do just that. I develop the Duality Squared model as a structurational conceptualization of the dualities constructing Internet governance.

Duality of Policy Making

A key element of policy-making discourse as social practice is the relationship it encapsulates between the agency of the policy makers and the social structures that both limit and enable that agency. This is the duality of the policy-making process. Structuration theory (Giddens, 1984) helps conceptualize links between the agency of individual actors and social structures, which the actors reify or alter through their mundane actions. It offers a language to describe the kind of messy constructs that come under the umbrella of information and Internet governance as constitutive processes.

Two core elements of structuration theory are structures and systems. Contrary to the traditional view of structure as an external factor constraining the agency (constructivism), structure in structuration theory is at least partially an internal attribute of the agent, which represents possibilities depicted in human practice and in the agents' memory. Giddens (1984) referred to it as "structural order of transformative relations," which exhibits "structural properties," that is, rules and resources that allow "binding of time-space in social systems" (p. 17). On the one hand, he described structural properties as the rules and procedures of action that are deeply rooted in our tacit practical consciousness. On the other hand, he viewed them as resources and power, or as the ability of agents to exercise their "transformative capacity" (Kaspersen, 2000, p. 42). Structures can be observed primarily through practice, such as adoption of information technology in organizational (Orlikowski, 2000) or other settings.

Unlike structures, social systems can be viewed as more explicit manifestations of structural relations (Giddens, 1984). They are the "relations between actors or collectives that are organized as regularized social practices and continually produced and reproduced" (Kaspersen, 2000, p. 45). Thus, law and policy are social systems, as are public transportation systems or any other explicitly organized relationship within a society. Social systems are the formalized or institutionalized versions of actual or desired routines of social practice. This conceptualization supports DeNardis's (2009) argument about technical protocols being a form of public policy insofar as they encapsulate ideas about freedom of expression, privacy, and so on.

Interacting with structures and systems are knowledgeable agents, who are purposeful and intentional in their actions and who can reflexively monitor their behavior and rationalize their actions (Giddens, 1984). In the context of policy making, discursive reflexivity—the ability of the agents to reflect on their and others' behavior and explicitly express their knowledge—is particularly interesting. The process of policy making is a process of discursive reflexivity deliberately aimed at altering the behavior of actors in society. Through discourse, the policy makers affect the public, but in doing so, they also affect the policy-making process itself. Any policy-making process is a system of making decisions that affects the public, and with each decision, policy makers reify the system's structural base regardless of the content of each decision. In Internet governance, this aspect is particularly salient, because institutionalization of Internet-related policy-making processes is at the heart of the debate. Thus the various processes of developing policy for the Internet reify the emerging structures of Internet governance.

The elements of the theory of structuration—primarily structures, agents, and systems—are inherently tied together and mutually influential. This leads to the central concept in Giddens's theory—the duality of structure—which suggests that the structure is both the medium and the outcome. As such, contrary to the traditional notion of structure, it is not a steady, external factor that limits agency but a rather constantly changing component that can both limit and enable agency and that is continuously challenged through practice.

Giddens (1979, 1984) described three groups of structures that explain the constitution of society. Structures of signification operate through framing or through interpretative schemes and involve the taken-for-granted knowledge assumed to be possessed by competent members of the society. These structures are used to identify typical acts, situations, and motives in a sustainable interaction. Through this interactional skill, which is essentially communicative, agents also recognize the intended and unintended meanings of acts.

Structures of legitimation operate through modality of norms (or rules) based on rights and obligations. If frames are used to identify acts, norms are used to assess the appropriateness of those acts. This in turn constitutes the duality of normative structures, because agents interpret normative structures, and each normative assessment has an array of behaviors it can evoke. As such, acceptance of norms is based on pragmatic assessment of normative and institutional alternatives. In other words, the agents have room "to produce a normative order as an ongoing practical accomplishment" (McLennan, 1997, p. 355).

Structures of domination operate through mobilization of power resources, allowing agents to secure their interpretation and normative claims in light of potential opposition from others. Such resources include organizational hierarchies, technical expertise, timetables and schedules, or anything that exerts control over the time and space dimensions of social life. Such resources also include interactional skills "involving high degrees of discursive penetration into the structures of signification and legitimation (such as the ability to argue successfully through the use of superior rhetorical skills or skills at normatively justifying one's position)" (McLennan, 1997, p. 356).

The process of policy making works through enacting these three types of structures across time and space, and it is also an explicit attempt to systemize a relationship among these three types of structures in a particular domain. This relationship is manifested in policy discourse as a form of social practice. For Internet governance, what matters is not only the substantive topics (e.g., management of Internet names and numbers) but also how decisions regarding these resources are made and how the correct or fair way to make these decisions is portrayed. A policy or a policy arrangement offers what Pinch and Bijker (1987) called a "rhetorical closure," meaning "whether the relevant social groups *see* the problem as being solved" (p. 44, emphasis in original).

However, policy and the process of policy making are never static. Building on Orlikowski's (2000) argument about the duality of technology, policy and policy-making processes are enacted through practice. As Giddens explained, "[h]uman actors are not only able to monitor their activities and those of others in the regularity of day-to-day conduct; they are also able to 'monitor that monitoring' in discursive consciousness" (Giddens, 1984, p. 29). The policy-making process, then, is an exercise in discursive reflexivity; it is a conscious attempt to encode norms and values in texts, an attempt to reflect, debate, and decide what is normative and what is not so it can be made explicit (see Braman, 2009b, 2011, for a specific Internet governance example). In this context, we see the policy-making and policy-debating spaces as the sites where agency is explicitly exercised and where structures of decision making are crafted.

As a discursive space, a forum that is explicitly dedicated to policy deliberation is an institutionalized form of modalities of structuration (Macintosh & Scapens, 1997). "Actors," according to Giddens (1984), "draw upon the modalities of structuration in the reproduction of systems of interaction, by the same token reconstituting their structural properties" (p. 28). Figure 3.1, reprinted from Giddens (1984, p. 29), represents the duality of structure as interconnectedness between the structures and their practice, practices that are often institutionalized in organizational settings. A nonbinding policy deliberation forum, for example, formally focuses on structures of signification, but those "always have to be grasped in connection with domination and legitimation" (Giddens, 1984, p. 31).

A policy discursive space, as primarily a modality of interpretive scheme, exists as a reification of structures of domination and legitimation. At the same time, it reproduces and reconstructs these structures through policy discourse as a social structure. More generally, according to Giddens (1984), "[w]hen social systems are conceived of primarily from the point of view of the 'social object', the emphasis is placed on the pervasive influence of a normatively coordinated legitimate order as an overall determinant of or 'programmer' of social conduct" (p. 30).

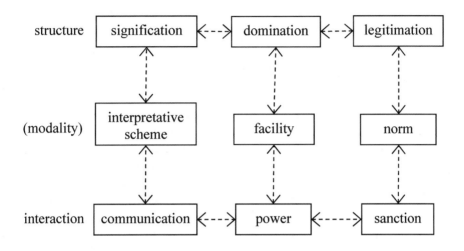

Figure 3.1: Dimensions of the Duality of Structure

Building on the notion of duality, I propose a conceptual framework for explaining the relationship between the process and the outcome of policy making. My conceptualization builds on the work of Orlikowski (1992, p. 410) and complements her modeling of the duality of technology. The duality of the policy-making model accounts for four types of influences among policy makers as agents, policy as a social system, and the context of policy making, which includes other social structures where the policy makers operate and the policy is being implemented. It views policy as both an outcome of human activity, such as international policy debates and negotiation, and a factor that facilitates and constrains policy-making activity through the existing structures of signification, legitimation, and domination. It accounts for the structural conditions of policy making, such as national and institutional identities, perception of technology, organizational settings of the debate, and so on, and at the same time it acknowledges the influences of implementation of policy on those and other social structures.

Viewing policy-making or governance processes through the lens of structuration theory highlights the role of policy discourse—or structures of signification in shaping the way we, as a society, come to think about information and communication technologies and their social roles. In this view, policy debates constitute instances of deliberative attempts to produce social systems through discursive reflection on competing social structures as manifested by the various stakeholders. As previously noted, in the case of information and communication policy, the social systems in question deal with socially constitutive powers, which are central to the processes of challenging and reproducing social structures (Banks & Riley, 1993; Braman, 2009a; Leeuwis, 1993).

The work of Orlikowski (1992) and others (e.g., Borg, 1999; Leeuwis, 1993) helps us to see how the argument about the duality of technology can be extended to information and communication technology policy. Similar to the creation of technology itself, technological policy is deliberately and consciously constructed by actors (policy makers) working in a given social context. However, policy is also socially constructed outside of that particular context through the different meanings actors (the public) attach to the technology and the various interpretations of the technological policy they emphasize and utilize in their daily lives. Thus, the process of constructing media, information, and communication technology involves both the designers and the users—all of them translate policy into practice.

Duality Squared

Pulling together the two notions of duality—that of policy and that of technology—offers a comprehensive conceptual framework for understanding the dynamics of Internet governance. I label this the Duality Squared model. Introducing this model requires one last conceptual exercise. We should note that notions of structures of signification, legitimation, and to a degree domination, are inherently communicative; it is through communication of and about social structures that human agents exercise their power resources (Bourdieu, 1991; Fairclough, 2001). Yet, although policy makers, especially in the field of information and communication policy, are explicitly involved in negotiating those structures and their relationships, ordinary citizens, who are not directly involved in policy debates, enact those structures through communication processes. To reiterate, "issues involving information and communication define the categories themselves and the relations enabled or permitted within and between them" (Braman, 2009a, p. 19). Since much of contemporary communication is mediated through technology, the process of negotiating the meaning of that technology defines social structures and is an influential factor in the constitution of society.

To describe the duality of technology within the Duality Squared model, one needs to focus on artifacts that constitute our mundane media environments. Building on Orlikowski (1992), we can still describe information and communication technology as both a product and a medium of human action, but at a macro-level, beyond the scope of a single organization. Here, technology as a medium is where structures of signification, legitimation, and domination are enacted and through which power resources are exercised. In turn, mundane uses of technology occur under social structural conditions of interaction with technology, such as cultural norms and perceptions of technology. Finally, there are social structural consequences of interaction with technology, such as exposure to alternative discourse, new venues for creative expression, or lower cost of collective action.

Linking this interpretation of the duality of technology with the structurational model of policy-making produces the Duality Squared model depicted in Figure 3.2. In addition to the relationships presented earlier, this model also includes policy as an outcome of human activity, such as international policy debates and negotiation, and as a factor that facilitates and constrains policy-making activity through the existing structures of signification, legitimation, and domination. It also accounts for structural conditions of policy making, such as nation and institutional identities, perception of technology, and so on, and captures the influences of implementation of policy on other social structures.

The governance processes of information and communication technologies constitute two mutually reinforcing dualities—thus duality squared. On one facet of this duality, policy makers react to unintended consequences for social structures and institutions created by diffusion and adoption of new technologies; at the same time, they set the agenda and provide guidance for future technological developments that impact social structures and institutions. On the other facet, while working on policy and regulations that mediate our abilities to communicate through technology, policy makers are acting within the limitations of the same social structures and institutions that are being influenced. Unfortunately, the two-dimensional representation used in Figure 3.2 does not adequately represent the complexity of the model. We must bear in mind that policy makers and policy discussants, who are the primary actors examined in this framework, are also human agents who interact with both the social structures and communication technology.

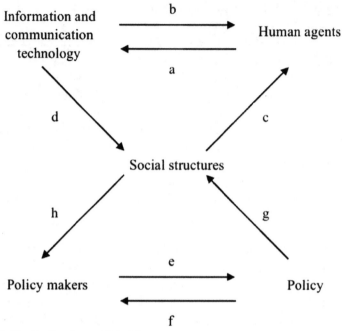

Figure 3.2: The Duality Squared Model

(a) Information and communication technology as a product of human action.

(b) Information and communication technology as a medium of human action, specifically a medium where structures of signification, legitimation, and domination are enacted and through which power resources are exercised.

(c) Social structural conditions of interaction with technology, such as cultural norms and perceptions of technology.

(d) Social structural consequences of interaction with technology, such as exposure to alternative discourse.

(e) Policy as an outcome of human activity, such as international policy debates and negotiation.

(f) Policy as a factor that facilitates and constrains policy-making activity through the existing structures of signification, legitimation, and domination.

(g) Structural conditions of policy making, such as nation and institutional identities, perception of technology, etc.

(h) Influences of implementation of policy on other social structures.

The Duality Squared model offers a flexible framework that can be applied to different domains of technology governance and policy making. For example, this framework would be particularly interesting and suitable for application in the sphere of Internet governance due to the high complexity and unique position of this sphere in terms of informational power. The model allows us to acknowledge that policy deliberation spaces are but one layer of Internet governance decision making; in particular, decisions are also made in other settings, such as the corporate world or communities of tech activists. The model also brings to the forefront the time- and space-related contexts of policy deliberation; this is an important aspect, because once developed and made public, policy discourse tends to become reified and institutionalized (as laws, regulations, standards, programs, etc.), thus losing its connection with the human agents that constructed it or gave it meaning; as such, policy discourse can come to appear part of the objective, structural properties of the society.

On the one hand, the Duality Squared model is general enough to allow discussion of broad social issues, such as those feeding the agenda of policy makers. By placing communication as the social activity at the center of our discussion, the Duality Squared model accounts for two substantively different yet mutually dependent relationships—one focused on policy makers and the other on human agents not directly involved in policy debates. On the other hand, the Duality Squared model is relatively specific and captures relationships that are inherently communicative and can be applied to a particular policy-making or policy-discussing setting focused on a specific information and communication technology (e.g., SOPA/PIPA, net-neutrality, multistakeholderism, etc.).

Concluding Remarks on Structuration of Internet Governance

The proposed structurational view of information policy making is a step toward a comprehensive conceptual framework of information governance through regulation of technologies that manage its flow. The emphasis on the communicative nature of enactment of structures of signification, legitimation, and domination further blurs the distinction between policy making and governance as determinants of questions to be asked. Instead, viewing both activities as exercises in discursive reflexivity allows asking comparative questions about the potential impact of binding and nonbinding, private and public, technical and social policy discussions on social structures and on governmentality. In other words, this may be a particularly suitable framework for the study of bottom-up and multistakeholder processes such as Internet governance.

The SOPA/PIPA example that opened this chapter is a good illustration of Duality Squared in action. The Duality Squared model reveals that the divide between the East Coast and West Coast codes goes beyond the literary meaning of code as two types of end product. Instead, the gap is about how those who chose to engage in the SOPA/PIPA debate used and perceived the Internet in fundamentally different ways. For policy makers in Washington, DC, for technology designers in Palo Alto, California, and for Internet users in the United States and elsewhere, the Internet evokes different modalities of structuration. Starting with the most fundamentally different views of the Internet as either a vehicle of commerce or a vehicle of creative work and free speech, through debate about the legal and technical facilities for carrying out proposed regulation, the standoff on January 18, 2012, suggests that governance of the Internet requires consent, or at least a commonly shared understanding of the Internet, by the various actors engaged in its shaping through regulation, design, and use.

The Duality Squared model is neither in the positivist sense nor a critical theory offering a normative judgment. It is a prism helping form questions about the dynamics of policy-making processes and the way they may alter social structures pertaining to communication. For example: How does policy establish meaning and norms of technology and at the same time reify assumptions about technology? How are previously nonnormative views made normative in the process of policy deliberation? What forces lead to systematic obfuscation of what may have been considered normative? Importantly, viewing policy making as a duality also allows us to ask questions about the actual agency of the policy makers: How do policy makers act as carriers of normative structures across different fora, geographic locations, and institutional settings? How often do public policy makers actually reflect on and rationalize activities and meanings that have already become commonplace, or do they accept and embrace meanings offered to them by private actors? What role do the structural properties of the policy-making process itself play, compared to the individual attributes of the agents in terms of their interpretation of priorities, opportunities, and constraints?

The Duality Squared model offers a conceptual map for a researcher trying to unpack the power dynamics that shape information policy, technology, and practices. The model tries to provide a unified framework that accounts for the multiplicity of factors in play, focuses on the dynamics of interaction, and steers away from the a dualistic view of social relations. The danger, of course, is a model that is too generic with limited explanatory power. Yet, as demonstrated by Orlikowski's (1992) and DeSanctis and Poole's (1994) adaptations of the theory of structuration, this conceptualization of duality is flexible enough to be applied at various levels of analysis and specificity.

References

Banks, S. P., & Riley, P. (1993). Structuration theory as an ontology for communication research. *Communication Yearbook, 16,* 167–196.

Battelle, J. (2012, January 19). On the problem of money, politics, and SOPA [Blog]. Retrieved February 27, 2015, from http://battellemedia.com/archives/2012/01/on-the-problem-of-money-politics-and-sopa.php

Borg, K. (1999). The "chauffeur problem" in the early auto era: Structuration theory and the users of technology. *Technology and Culture, 40*(4), 797.

Bourdieu, P. (1991). *Language and symbolic power* (G. Reymond & M. Adamson, Trans.). Cambridge, MA: Harvard University Press.

Braman, S. (2009a). *Change of state: Information, policy, and power.* Cambridge, MA: The MIT Press.

Braman, S. (2009b). Internet RFCs as social policy: Network design from a regulatory perspective. *Proceedings of the American Society for Information Science and Technology, 46*(1), 1–29.

Braman, S. (2011). The framing years: Policy fundamentals in the Internet design process, 1969–1979. *The Information Society, 27*(5), 295–310.

Bridy, A. (2012). Copyright policymaking as procedural democratic process: A discourse-theoretic perspective on ACTA, SOPA, and PIPA. *Cardozo Arts & Enterntainment Law Journal, 30*(2), 153–164.

DeNardis, L. (2009). Protocol politics: The globalization of internet governance. Cambridge, MA: The MIT Press.

DeNardis, L. (2010, September 17). The emerging field of internet governance. *Yale Information Society Project.* Retrieved February 27, 2015, from http://papers.ssrn.com/sol3/papers.cfm?abstract_id=1678343

DeSanctis, G., & Poole, M. S. (1994). Capturing the complexity in advanced technology use: Adaptive structuration theory. *Organization Science, 5*(2), 121–147.

Fairclough, N. (2001). *Language and power.* New York, NY: Longman.

Fischer, F., & Forester, J. (Eds.). (1993). *The argumentative turn in policy analysis and planning.* Durham, NC: Duke University Press.

Flyverbom, M. (2011). *The power of networks: Organizing the global politics of the internet.* Northampton, MA: Edward Elgar.

Galloway, A. R. (2006). Protocols vs. institutionalization. In W. H. K. Chun & T. W. Keenan (Eds.), *New media, old media: A history and theory reader* (pp. 187–198). New York, NY: Routledge.

Genieys, W., & Smyrl, M. (2008). *Elites, ideas, and the evolution of public policy.* New York, NY: Palgrave Macmillan.

Giddens, A. (1979). *Central problems in social theory: Action, structure, and contradiction in social analysis.* Berkeley: University of California Press.

Giddens, A. (1984). *The constitution of society.* Berkeley: University of California Press.

Hart, J. A. (2011). Information and communications technologies and power. In S. Costigan (Ed.), *Technology and international affairs* (pp. 203–214). Surrey, UK: Ashgate.

Kaspersen, L. B. (2000). *Anthony Giddens: An introduction to a social theorist.* New York, NY: Wiley.

Leeuwis, C. (1993). Towards a sociological conceptualization of communication in extension science: On Giddens, Habermas and computer-based communication technologies in Dutch agriculture. *Sociologia Ruralis, 33*(2), 281–305.

Lessig, L. (2006). *Code. Version 2.0.* New York, NY: Basic Books.

Macintosh, N. B., & Scapens, R. W. (1997). Structuration theory in management and accounting. In C. G. A. Bryant & D. Jary (Eds.), *Anthony Giddens: Critical assessments* (Vol. 15, pp. 455–77). New York, NY: Routledge.

Mathiason, J. (2009). *Internet governance: The new frontier of global institutions.* New York, NY: Routledge.

McLennan, G. (1997). Critical or positive theory? A comment on the status of Anthony Giddens' social theory. In C. G. A. Bryant & D. Jary (Eds.), *Anthony Giddens: Critical assessments* (pp. 318–326). New York, NY: Routledge.

Mueller, M. L. (2010). *Networks and states: The global politics of internet governance.* Cambridge, MA: The MIT Press.

Mueller, M. L., Kuehn, A., & Santoso, S. M. (2012). Policing the network: Using DPI for copyright enforcement. *Surveillance & Society, 9*(4), 348–364.

Netburn, D. (2012, January 19). Wikipedia: SOPA protest led 8 million to look up reps in Congress. *Los Angeles Times.* Retrieved February 27, 2015, from http://latimesblogs. latimes.com/technology/2012/01/wikipedia-sopa-blackout-congressional-representatives.html

Orlikowski, W. J. (1992). The duality of technology: Rethinking the concept of technology in organizations. *Organization Science, 3*(3), 398–427.

Orlikowski, W. J. (2000). Using technology and constituting structures: A practice lens for studying technology in organizations. *Organization Science, 11*(4), 404–428.

Pinch, T., & Bijker, W. E. (1987). The social construction of facts and artifacts. In T. P. Hughes & T. Pinch (Eds.), *The social construction of technological systems* (pp. 17–50). Cambridge, MA: The MIT Press.

Samuels, J., & McSherry, C. (2012, January 18). Thank you, Internet! And the fight continues. Retrieved February 27, 2015, from https://www.eff.org/deeplinks/2012/01/thank-you-internet-and-fight-continues

Singh, J. P. (2009). Multilateral approaches to deliberating internet governance. *Policy & Internet, 1*(1), 91–111.

Produsing the Hidden: Darknet Consummativities

Jeremy Hunsinger

Consummativity is a mode of the labor of consumption. It is both its measure and the system of needs driving consumption itself as constituted through and constituting the subject's desires (Baudrillard, 1981; Dant, 1999). As a field of distributed subjectivity, consumption is one mode through which we coconstruct ourselves and our relation to things. Darknets are clearly participating in this field of consumption, as is the Internet in which darknets exist. Consummativity is the labor of the desires that coconstitutes the subject of the produser in relation to his or her consumption of the project (Bruns, 2006, 2008a, 2008b). In the case of darknets, that consumption is of time, knowledge, and economic value involved in produsing darknets.

Consummativity is both the identification and alienation of the reflexive awareness of the subject's desires/needs. On darknets, these desires/needs are coconstructed to run the gamut from the simplest curiosity to political organizing to representations of illicit activity to the activities themselves. The alienation and identification of desires function much like the creation of the identity of producer/user or produser: the quasiobject that is produced is always partly the produser's alienated identity (Bruns, 2008a, 2008b; Gehl, 2014; Herman, 2013; Latour, 1993). This process produces the capacity to realize the values in the darknets, including the values found in the relationship to the material presented therein. Given the proclivities of the representations of goods and services on darknets, this chapter examines how consummativity functions to create and recreate the darknet, both as a thing and as an idea.

Fundamentally, my argument resists the simple construction of criminality and exceptionalism found in popular press understandings of darknets, in favor of an understanding that although the criminal element does exist, the broader nature of consummativity in darknets is about knowledge, information provision, and economic value creation. In short, I argue that the representation of material and nonmaterial goods on darknets creates a consummative zone, allowing the subject to be reconstituted with the awareness of an extended horizon (Buck-Morss, 2002; Hunsinger, 2011; O'Tuathail, 1994). This zone, with its horizons, fetishizes elements of the consummativities such as hiddenness.

On Darknets

Darknets, darkwebs, and the related conceptual entities described in this chapter are Internet-based information systems that limit and redirect access into intentionally hidden or very difficult-to-enter areas. I present them primarily as a conceptual entity because they are contiguous with the Internet as a whole, but they are thought of differently; although darknets have a referent set of things in the world, those things are fluid and changing in a broad ecology of information services. The borders between darknets and the consumer Internet are similarly fluid, and some of the arguments I make here can apply to both, because darknets are a subset of the Internet. A more technical understanding of the type of darknet that this chapter discusses is that darknets are securitized Internet networks operating over existing networks through encrypted traffic on those networks, or—increasingly— they are mixes of those networks and either planned or ad hoc mesh networks. Mesh networks are computer-to-computer networks that route messages through many computers acting as a network and frequently operate outside of the consumer Internet.

A key technosocial point of a darknet is that it is either technologically or socially designed to be trusted or trustable. That is, users should be able to trust the securitized nature of the darknet with their activities, although sometimes that security will fail. Various modalities of darknets predated the Internet; they existed on bulletin board systems, and in all likelihood they will exist on almost any securable network where its produsers can create trusted connections.

Darknets can be thought of as infrastructural elements of populations of producers/users that have interfaces that those produsers use. Because these interfaces are hidden from the consumer Internet through encryption and other means, technical or community expertise is required to find and use these interfaces. Such expertise is recognized by scholars who study expertise as a necessary element to accessing and controlling infrastructures such as sewer systems, medical information systems, or traffic control systems (Star, 1999, 2002). As infrastructures, they are not only intentionally hidden from common users of the consumer Internet, but they also disappear from our critical horizon until a failure brings them to our attention (Hunsinger, 2005).

Examples of darknets include the hidden wiki and other services in Tor .onion services, I2P services, various secured FTP and encrypted P2P services, and Freenet services. Darknets arise for any number of reasons, but generally they are motivated by issues of surveillance, privacy, security, and economic value, which are rooted in the needs and desires of their produsers. Economic value is a significant motivator in a capitalist society, but beyond that it is complicated, because economic gain might occur in a myriad of ways. For instance, if you increase your trust in your network by relaying more and valuable information, you might gain social capital, which could generate economic gain. Or you could download a series of costly digital objects from trusted friends, which might also increase your economic value. Those two actions are likely the most common modes of economic value creation on darknets; they happen on all darknets because trust and sharing are central to their social operation.

Criminal economic gain also occurs via darknet systems (as it does on the consumer Internet), but other than dealing in digital (rather than tangible) goods, darknets function as do any other marketing arena for criminal activity. I recognize that this is a part of darknet activity and a motivation for some users of darknets, but in this chapter I do not deeply engage with the mythogenesis of crime on darknets. For instance, the hidden wiki on .onion has advertisements for money laundering, child pornography, murder-for-hire, bomb making, drug trading, contraband selling, the sale of extremely cheap and likely fraudulent items, and many other things. However, the representation of the criminal activity on darknets is primarily mythogenetic. That is, the information represented on the entering pages is meant to scare populations away if they may have stumbled upon the place while merely exploring. It is meant to create or ground those new users' mythos of darknets. The important thing is to make populations believe these activities are happening, so they stay away. This is part of the practice of camouflage and hiding darknets so that casual users are dissuaded from participating.

Many produsers also engage in political activity via darknets, which is reason to seek privacy and avoid surveillance. The political activity on darknets is wide ranging, from activities such as the group Anonymous, to communist and libertarian groups, to animal liberation activities, and international political organizing (Coleman, 2014). Although some of these activities are clearly nonviolent and to some extent not really happening at all, others, such as the Anonymous and WikiLeaks activities in darknets, have had real implications for governments and their subpolitical proxies such as corporations (Beck, 1998, 2000). Much like criminal activities, political activities on the darknets are mythogenetic and contribute to the myths about darknets. This element of mythogenesis and promotion of the existence of noncriminal political and cultural activities on the darknets are part of how darknets legitimate their existence and continue to be portrayed as necessary spaces for some produsers.

One final mode of use of darknets—and as I have mentioned, it is likely the largest—is that of sharing or trading knowledge. Produsers go to the darknets to learn things that other people might consider contraband knowledge. This knowledge might be hard to find or even censored on the publicly accessible consumer Internet, but some is concentrated and easy to find on darknets. This is not necessarily knowledge of a criminal nature; the darknet has book repositories, scientific article repositories, e-book trading zones, and many other systems for sharing knowledge related to a wide variety of topics such as political theory, gender studies, physics, chemistry, and engineering. Some of this knowledge is shared beyond the enclosures of its original copyright or otherwise outside of its licenses, but some material is shared legitimately. Although the presence of this material does not legiti-mize the practice of sharing the rest of the material, it does to some extent provide imagined efficiencies for the produsers of darknets trying to access that material. If these efficiencies might become empowerments to people in disadvantaged situations (e.g., facing censorship), darknets gain legitimacy through the presence of this information. This legitimacy is similar to the modes of legitimation of the consumer Internet, in which knowledge sharing is dominant. Knowledge sharing was one of the primary ideas for the original Internet and hypertextual systems (Cailliau & Gillies, 2000; Nelson, 1987; Raymond, 2001).

The activities produsers pursue on the darknets are both real and hyper-real. The activities become hyperreal when—even if they do not actually exist—their existence or nonexistence is indistinguishable from their semiotic existence. Murder-for-hire on darknets is an example. It may or may not exist, but enough people act as though it does to make it hyperreal. Other hyperreal activities include global political revolution and information about space-based alien life. In contrast to the hyperreal, knowledge sharing, political activity, and some forms of criminality are certainly real activities. Both the real and hyperreal help legitimate the existence of darknets in the eyes of some of their produsers as ways to both encourage some beneficial economic and political activities and enclose some potentially nonbeneficial activities. In both cases, darknets exist in part to hide material from the prying eyes of surveillant assemblages such as corporations, governments, and neighborhoods, while allowing for the produsers to coconstruct the functions of exceptionalism or eliteness that might encourage them to continue their pursuits in these zones (Haggerty & Ericson, 2000).

Expertise and exceptionalism relate back to the hiddenness of darknets, because the knowledge of and access to these darknets are some of the motivations people have for pursuing their interests through darknets. People who know their situatedness in the surveillant assemblage differentiate from others who do not. Knowing the security and trust in one's darknet is also a key point for these systems of expertise and their relevant communities, because ignorance of those details will cause failure of one's own security and privacy. Darknets center on knowledge, specifically the relation between that knowledge and the hiddenness as it relates to their produser communities and the technical expertise of privacy and security of networks. Those relations of desire/need/knowledge define the system of consummativities of darknet produsers.

On Consummativities

The hiddenness of darknets is one of the consummativities that its produsers create within the system of relations that constructs the darknets, the knowledges of darknets, and the desires/needs enveloping darknets, thus constructing its produsers as subjects in relation to darknets. The mediation and semiosis of the object between the producer and user transform the relationship between the thing and its communities (Bruns, 2006, 2008a, 2008b). The idea of a produser requires consideration of consummativity and thus the semiotics of desires/needs within integrated world capitalism (Baudrillard, 1981; Guattari, 1989). Integrated world capitalism is a capitalism that centers less on the construction of nation-state and capital actors than on the construction of good capitalist subjects, with desires/needs that can be satisfied within structures of control and semiotic production rather than in systems of production of goods and services (Guattari, 1989). Although darknets do distribute goods and services, these consumables are also semiotically produced within the system of consummativity that exists within integrated world capitalism.

Consummativity is a concept originally developed by Baudrillard to discuss what he described as a political economy of the sign, where the idea of the sign is at once a normal semiotic sign and a signifier of economic value. Indeed, the base idea is that the capitalistic construction of value always intervenes between the signs and the signified (Baudrillard, 1981). This is a double meaning, though, because not only are the signs themselves imbued with capital and the excess of a general economy, but the conveyance of the sign also implies capital of and through the subject/object that relates to it (Bataille, 1988; Baudrillard, 1975, 1981, 1993, 1998). With that double valence of meaning in mind, consummativity is a concept that is developed as a parallel relation to productivity: As productivity is to production, so consummativity is to consumption. This relation defines the relationship of productivity to production as existing in tension with the subject in terms of measurability—like Taylorism but as a cultural construct of neoliberal corporatist consumption, so that we consume productivity as much as the products themselves.

Consummativity is an underdefined process that understands the subject as a person with underdefined needs; because the needs are underdefined, integrated world capitalism as a series of processes of neoliberal corporate consumption can help the subjects define their needs and thus their satisfactions. But the claim is less that these processes help define the subject's needs than that they help redefine the subject as a produser of needs, which phenomenologically seems undeniable in the context of capitalism:

> And just as the fundamental concept of this system is not, strictly speaking, that of production, but of productivity (labor and production disengage themselves from all ritual, religious, and subjective connotations to enter the historical process of rationalization); so one must speak not of consumption, but of consummativity: even if the process is far from being as rationalized as that of production, the parallel tendency is to move from subjective, contingent, concrete enjoyment to an indefinite calculus of growth rooted in the abstraction of needs, on which the system this time imposes its coherence—a coherence that it literally produces as a by-product of its productivity. (Baudrillard, 1981, p. 83)

The system of needs and its production is as such coherent, but only because it is produced within the abstraction of these needs and their reproduction as engines of growth in a capitalist system. That is, the system of needs that we construct within our semiotic world is perpetually capitalistic, but it is also designed to satisfy our needs even if we imagine ourselves noncapitalistic, as do some produsers of darknets. The construction of the system of needs is the foundation for consummativity. The labors of our production of needs and the labors of these needs' consumption operate through our distributed subjectivities from which we perpetually abstract needs. As we abstract needs, we also abstract the labors for their re/creation. Both abstractive labors are aspects of the other, and each process may be what Baudrillard called unlimited abstraction, which, as part of the capitalist-semiotic system, generalizes both the needs and their labors until the system is sustained without subjects. The subjects as abstractions of desire/needs become hyperrealities. However, darknets are not, as of yet, hyperreal constructs—although clearly some of their consummativities may be. Darknets, like the Internet, are real systems with real produsers, but we must consider how the existence of darknets constructs their users and their respective consummativities. As Baudrillard (1981) said:

> . . . it is consummativity that is a structural mode of productivity. On this point, nothing has really changed in the historical passage from an emphasis on "vital" needs to "cultural" needs, or "primary" needs to "secondary" ones. The slave's only assurance that he would eat was that the system needed slaves to work. The only chance that the modern citizen may have to see his "cultural" needs satisfied lies in the fact that the system needs his needs, and that the individual is no longer content just to eat. In other words, if there had been, for the order of production, any means whatever of assuring the survival of the anterior mode of brutal exploitation, there would have never been much question of needs. Needs are curbed as much as possible. But when it proves necessary, they are instigated as a means of repression. (p. 84)

As the simple relationship of slave to work implied food to the slave, so too the complex relationship between darknets and their produsers implies that there must be needs or desires that are unmet or unrealized. These may be exploited, and if not exploitable, they must be curbed or repressed. These needs must be realized as consummativities for them to be exploitable, and if so, then they must be realized to some extent in both the desires and their fulfillment. Darknets have semiotic relations indicating their consummative relations to their produsers. However, those semiotic relations may be hidden from the produser's perception because the interpretive horizon might not perceive the structural system in which the signs and semiosis operate. Thus the hiddenness is due in part to the capacity of the population to ignore or perhaps deny their relations in integrated world capitalism. Dant (1999) confirmed this reading of Baudrillard:

> In this system of objects as sign values and their exchange that Baudrillard terms 'consummativity' (1981:83), a dynamic of capitalist society that he juxtaposes to productivity. Consummativity is the system of needs for objects imposed on individual consumers; it includes their need for choice. Needs cannot be derived from a humanistic notion of the free, unalienated, asocial individual driven by craving or pleasure or even by some essential needs. (p. 50)

Dant pointed out the necessity of the group or organic organization for the consummative organization of the needs of humanity. This is also true of the consummativities of darknets, because without the social, political, and economic organizations around them, darknets could not exist. Like all of the consumer Internet, the social, political, economic, and even bureaucratic organizations of darknets are predicated on prior forms (Cerf & Kahn, 2004; Hunsinger, 2013), which also provide the context and organizational background for the consummativities. Produsers of darknets learn to need certain forms of social, political, economic, and bureaucratic systems, learn to need certain types of control, and learn to develop their interests, desires, needs, and knowledges within those frameworks.

Among comfortable prior forms, the corporate market is well-established in darknets. There have always been consummativities surrounding the activities currently advertised and found in darknets, whether alternative political arenas or complete wild zones containing criminalized or censored objects, goods, or services (Buck-Morss, 2002; Luke, 1995; O'Tuathail, 1994). Goods that resolve the consummativities currently resolved by darknets have always been accessible in the circuits of consumption, but now on the darknets, a produser of marketable commodities can engage in that activity in an anonymous or pseudoanonymous manner. This transformation of the mode of engagement engages a need for something not present in all markets: the need for privacy in the transaction. The assumption of the trust of the darknet's security and privacy is an order of magnitude above that of the consumer Internet. In the past, this need for darknet technologies was a marketable commodity itself, but with open source technologies, costs have lowered, and it has become a feature of the postidentity Internet marketplace.

We do not necessarily need identity for markets to operate; we need only consummativity and its relation to the flexible subject of integrated world capitalism, where the public and private have collapsed into cooperant perceptive/semiotic valences of the subject. The circuits of identity are part of the reproduction of organizational forms, as noted by Luke (1999):

> This co-modifying circuit of commodified reproduction elaborates the essential logic of "consummativity," which anchors this entire system. Instead of maintaining the irreducible tension between the public and private spheres that liberal economic and legal theory accept as true to accept the individual contingency of rational living, the public and private have collapsed into co-modifying circuits of identity all across the technosphere in the code systems of corporate-managed consummativity. (p. 71)

This collapse of the public/private into the articulation of desires/needs of the flexible subject is part of the processes of consummativity involved in the articulations of expert systems of knowledge. Formerly only profession-als could collapse one's public/private identity into articulations of need. Doctors, lawyers, and government officials could know the relationships among elements of one's self that one might seek to keep separate (Illich, 1977). In darknets and similar technologies, the subject can and must collapse his or her desires/needs into the anonymized subject of consump-tion. This effectively hides subjects' identities from the remediations of organizations of the Internet (Bolter & Grusin, 2000; Cerf & Kahn, 2004; Grusin, 2010).

That identities on darknets are hidden is central both to their appeal and to their related consummativities. We should not confuse their implicit hiddenness with security or privacy, because security and privacy also have consummativities that drive interest in darknets. One other significant consummativity relates to exceptionalism, elitism, and the constructions of expertise around knowledge of darknets, what I refer to elsewhere as the difference between the in and the within (Hunsinger, 2013). These consum-mativities are central to the construction of darknets, but it should be noted that they operate transversally across our organizational assemblages. Consummativities shift among levels of analysis, comparing, contrasting, and distributing the desires/needs of the populations that exist in different organizational arrangements, from ideologically derived individuals, to corporations, and to nation-states. At each level of organizational complexi-ty, consummativity operates and distributes consummative realities. This operation and distribution allow our desires/needs to be coconstructed into the productive asset of integrated world capitalism. This productive asset is what drives produsers to cocreate darknets to serve the perceived needs of their populations. These needs are implicit in the circulation of signs within the zones in which darknets exist and in the distributed subjectivities of the population at large. In darknets, produsers feel the need for certain psychic states and seek to fulfill the need by produsing the darknet and all of its attendant organizational constructs having both semiotic and mythogenetic operations.

As produsers of darknets realize their situatedness within these con-structs, their awareness of their relations within them becomes more obvious. They realize that they are coconstructing the environment by being consum-ers of darknet services, influencing design by selecting which modes of security and privacy-preserving technologies to use and choosing whether to be anonymous, among other darknet-oriented identity performances that demonstrate desires/needs. Produsers need to understand that although knowledge and other goods of various dubious and nondubious qualities exist on darknets, the darknets do not exist solely for their coconstruction of needs. Darknets form a wild zone of contested consummativities in which our desires/needs are continually coconstructed, without the possibility of returning to the nonexistent true or natural state. What we have is the circulation of signs of nature and truth, which then generates the individualized production of desires/needs (Dant, 2004).

Reincorporating Darknets and Their Consummativities

Darknets are exemplary of this production of individualized needs; they function mostly in relation to their circulation of signs as coconstitutive of the produser communities. People build darknets, but their histories do not tell the stories of their originations insofar as those histories help to generate the systems that coconstruct the desires/needs.

Darknets are the accident of what is called *mode 2* science production: the scientific production that produced encryption, hypertext, and internetworking technologies (Gibbons et al., 1994; Nowotny, Scott, & Gibbons, 2001; Virilio, 2000, 2007; Virilio & Lotringer, 2002). When we produced the applied knowledge making those three technologies possible, the logical next step was to develop encrypted, hypertextual, and distributed information systems. To use these systems, one needs to know how to find them and how to navigate them, and by implication, how to maintain their security.

The technologies that come together to form darknets have their own independent histories, but the know-how necessary to make them work together is embodied as expertise within populations of programmers, systems administrators and information security professionals (Abbate, 2000; Cailliau & Gillies, 2000). Securing information systems has been a long-term pursuit in many of these knowledge communities: securing information that was private to the enterprise or was required/desired to be secured by some private, legal, or corporate motivation. These securitizing pursuits express the desires/needs of those organizations and coconstruct the consummativities in which these knowledge communities participate.

The pregenitors of darknets (business, governmental, paramilitary, and military entities) have long been accustomed to enclosing knowledge to keep it from the public, creating a set of expectations and values around knowledge that prefigured the consummativities of darknets. As darknets began to be used to privatize knowledge and thus allow it to be more successfully marketed and capitalized, users of secured networks expected even more enclosure of knowledge. Such enclosure of knowledge arguably makes knowledge more marketable; it is one of the fundamental constructs of professions as described by Ivan Illich (1977), and of control society as described by Deleuze (1992).

The entities using these early darknets employed them quite effectively as a tool of control within neoliberal governance, providing network elites with access to knowledge, empowering them to act in ways that would in part be prefigured by the necessary construction of difference based on membership in the population that knows and has access to knowledge. Because they were so effective in controlling privatized information for capital purposes, these darknets evolved into significant investments for the institutions projecting darknets as power into the world. This investment drove more mode 2 research into the underlying technologies, publicizing the knowledges involved in creating darknets and the hardware and software that could allow them to be produced at the consumer level.

Once consumer-grade darknets became available, smaller communities and individuals could use darknet technologies, thus allowing a broader set of consummativities to be realized. Governments became interested in regulating darknets, but by then the technologies were already widely used by many different populations. The dispersal of technologies combined with the increased research funding boosted the capacities of these technologies through the pluralization and thus differentiation of projects. Eventually, anonymity projects such as Tor and Bitcoin came to fruition.

Governments and military establishments funded some of the new darknet technologies, creating a tension rooted in different understandings of the nature of darknets held by the technology funders and the technology users. This tension highlights the consummativities of the various populations and institutions involved in darknets by showing that noninstitutional darknet users—people who use darknets for a variety of personal reasons—don't always care deeply about who created or pays for their darknet. This attitude is completely different from that of institutional users, who have parlayed darknets into economic advantage and thus seem to need to know and trust the creators more directly.

The Consummativities of Privacy and Security on Darknets

On one level, my argument is about privacy and surveillance because that is the center of the desires/needs around darknets and thus the consummativities of darknets. As noted previously, consummativities occur across all levels of organizational being, from individuals to nation-states and beyond. Two desires shared across several types of organizations are privacy and security; darknets exemplify one attempt to fulfill both.

Privacy is a difficult desire/needs complex to parse across all levels of organization. Although it is defined in law and somewhat defined in culture, privacy on the consumer Internet only exists if one has security. On the consumer Internet, without the knowledge of your security, you cannot be assured of privacy. You may construct privacy as a mode of resistance to surveillance as a concept, resistance to surveillance in real life, or for any set of desires and knowledges that are salient about surveillance to you. Some populations do not even resist surveillance. The mode of construction of privacy that those populations have is usually complicity. This complicity with surveillance is much like sousveillance (when we watch the watchers, or watch the surveillance) or even entertainment, such as the Surveillance Camera Players (SCP), because it simultaneously coconstructs and is nascently critical of the surveillance. The SCP put on theater for surveillance installations nominally to entertain those who are watching the monitors. However, surveillance even with manifestations of those modes of both complicity and nascent criticality is still subpolitically located across fields of legitimation, whether artistic or everyday life (Beck, 1998, 2000).

Internet surveillance is sometimes subpolitical because it represents the transference of the responsibility to govern Internet use from our governments, elected and otherwise, to nongovernmental entities. These institutions govern with our consent either in relation to the general will of a democratic public or as a leviathan (Hobbes, 1994; Hunsinger, 2013; Rousseau, 1997). Beyond the purported Rousseauian general will, corporations exist for a wide variety of reasons but primarily must serve their stockholders. Thus, even though this subpolitical realm may be legally constructed, it is frequently constructed within markets and related desire/need constructions of corporations and governments. These need constructions are consummative, existing primarily on the semiotic and interpretive levels until projected as power in the world. The desire/need constructions of corporations are the primary fields where the subpolitics of consummativity is manifested. These desire/need constructions vary significantly across corporations but can engage and/or parallel constructions of the public good. The desire/need for public goods is also consummative in nature; it is coconstructed across the populace through its own institutionalization. The coconstructed institutionalizations of the public good allow these corporations to represent legitimized actions in the eyes of the population and thus the government. Corporations perform innumerable actions to represent the appearance of operating for the public good; in representing that good, they are coconstituting consummativities. In short, Internet surveillance is justified as something that both governments and publics desire/need. However, the desire/need for surveillance also begets the need for resistance of surveillance. This resistance of surveillance is desired/needed across individuals, corporations, and governments, and it begets the desire/need for darknets.

Corporations as subpolitical actors are diffusers of government power as well as concentrators of surveillance and darknet technologies. That is to say, although there are hundreds or thousands of corporations acting politically in the world, few of those corporations concentrate control over surveillance and darknet technologies to create their profits. Thus as governmentalities distribute across corporations subpolitically, the power of these corporations has spaces and flows where it is concentrated and intensified (Burchell, Gordon, & Miller, 1991). Frequently, these spaces and flows of power are heavily branded, publicly identifiable corporations such as Facebook, or Rogers Media in Canada, but sometimes these corporations are very private. Whether public or private, the projections of power in regard to surveillance and darknet technologies are semiotic projections that demand readings and responses from populations. As populations read and respond, they coconstruct desire/need relations that will vary along a spectrum from open resistance using darknet tools to complete complicity using the consumer Internet and self-surveillance.

Because privacy and security are coconstructed as desire/need and public good, we fund research into technologies that support them and thus transfer wealth and power to corporations around those technologies. Even though these corporations are projecting identities that center on the public good, their projections tend to define the semiotic terrain, and thus they coconstruct the possible modes of interpretation. For example, are Google's and Facebook's understandings of privacy and surveillance the same? Or even the same as those within the cultures in which they operate? Surely that is impossible, but they must project their understandings into those cultures and to each other because the understanding of privacy and surveillance that they project enables their technology. This capacity that corporations have over our consummativities is possible in part due to our highly individualized society with diverse understandings. It is also partly due to a political imagination that limits the horizon of the possibility of unity and community in the face of corporate and government power. Culturally, our political imagination of our own individualism and the necessity of privacy as an individual or familial construct enables the justification of surveillance. These political imaginations and their horizons are elements of the system that develops our consummativities. As such, even our politics is a consumable semiotic good within integrated world capitalism (Guattari, 1989).

The limits of our political imagination are exemplified by the tendency to assume that privacy is only a right. But in our liberal society, there is an assumption that many are extremely wary of making about privacy: privacy may be a right, but it is also a consumer good. Because they are consumer goods as much as public goods, privacy and security also have consummativities. It is in part those consummativities that produsers recognize when they desire/need certain constructions of privacy/security in online environments. Although privacy is projected as a public good in a liberal society and economy, privacy in our neoliberal economy only has value where it is traded or consumed. Being a public good does not necessarily give it value; neither does its status as an acclaimed right. As a consumer good in a neoliberal economy, it gains value from trade and the labors put into it as it is consumed—its consummativities. The consummativities of privacy are those labors based on the perceived need of privacy and which are fundamentally founded in our desires, which are culturally, politically, and otherwise bound by various instrumentalities and rationalities.

The instrumentality of choice is a prime example. Many different sets of darknet technologies can do basically the same set of operations. One darknet will provide easier access to some actions, and another will better enable other actions, but the choice between darknets is one of satisfaction of semiotically coconstructed desires/needs and not one necessarily of actually being secure or private in the darknet-supported interactions. Consummativity only depends on the brand and one's capacity to choose among those brands, for whatever reasons and instrumentalities that one construes:

> Is it not true that consummativity, as a mechanism of power, depends on the cultivation of the body and what it does, allowing capital to extract time, energy, and labor from what the body does in production and consumption? As the American populists feared, the material coercion of consummativity is to be found within its authoritarian confinement of local choice and totalitarian denial of global alternatives. Consummativity is liberating, but only in the peculiar fashions that it will determine within its macroenvironments. (Luke, 1999, p. 230)

Thus we can begin to understand that consummativity is an encapsulating idea that also generates our rich life full of choices within the fashion system as the system of consumer choices based on brands, designs, and identitarian politics, or, alternatively, our life, full of choices, is based on stylistic rather than substantive differences. I do not want to demean darknet produsers with the claim that darknet technologies are merely fashion choices, but it should be clear that given the range of possibilities, fashion is one way to assert the exceptionality, the knowledge, and the sociality that are part of darknet populations. The security/privacy construct of integrated world capitalism is also one of those fashions of consumption. In that it is fashion, it is programmed by our desires/needs and has consummativities.

Darknet Consummativities

The consummativities of privacy and security as noted earlier relate to the knowledge of security and resistance to surveillance, among other laborious endeavors. Darknets are part of a global system of information provision. They are a vital part of that system because they provide avenues for populations to choose to be the same as certain kinds of people and to be different from other kinds of people. In making that choice, these populations will help to resolve some of their consummativities around surveillance and privacy. This act of choosing will certainly expose them to many kinds of information to which they might not want to be exposed. It will also give them a particular sense of their subjectivity, their knowledges, and their own expertise. Those transformations of subjectivity are coconstructed with the consummativities of darknets. They are the consummativities of a series of hidden populations, all choosing to hide on darknets for a desire/need that has been coconstructed through the semiotics of integrated world capitalism. The modes of being of these produsers of darknets all hinge upon those perceptions and those choices, which may be portrayed as natural or true choices for security or privacy but again are also coconstructing the produsers as darknet consumers. Even if they are acting politically or criminally in those environments, they are still produsing the consummativities of that politicality or criminality.

References

Abbate, J. (2000). *Inventing the internet.* Cambridge, MA: The MIT Press.

Bataille, G. (1988). *The accursed share: An essay on general economy.* New York, NY: Zone Books.

Baudrillard, J. (1975). *Mirror of production* (M. Poster, Trans.). St. Louis, MO: Telos Press.

Baudrillard, J. (1981). *For a critique of the political economy of the sign.* St. Louis, MO: Telos Press.

Baudrillard, J. (1993). *Symbolic exchange and death.* London, UK: Sage.

Baudrillard, J. (1998). *The consumer society: Myths and structures.* London, UK: Sage.

Beck, U. (1998). *Democracy without enemies.* London, UK: Polity Press.

Beck, U. (2000). *What is globalization?* London, UK: Polity Press.

Bolter, J. D., & Grusin, R. (2000). *Remediation: Understanding new media.* Cambridge, MA: The MIT Press.

Bruns, A. (2006). *Towards produsage: Futures for user-led content production.* Proceedings from Cultural Attitudes towards Communication and Technology, Tartu, Estonia.

Bruns, A. (2008a). *Blogs, Wikipedia, Second Life, and beyond: From production to produsage.* Bern, Switzerland: Peter Lang.

Bruns, A. (2008b). The future is user-led: The path towards widespread produsage. *Fibreculture, 11.* Retrieved February 25, 2015 from http://eleven.fibreculturejournal.org/fcj-066-the-future-is-user-led-the-path-towards-widespread-produsage/

Buck-Morss, S. (2002). A global public sphere? *Situation Analysis, 1,* 10–19.

Burchell, G., Gordon, C., & Miller, P. (1991). *The Foucault effect: Studies in governmentality.* Chicago, IL: The University of Chicago Press.

Cailliau, R., & Gillies, J. (2000). *How the Web was born: The story of the World Wide Web.* Oxford, UK: Oxford University Press.

Cerf, V., & Kahn, R. (2004). What is the Internet (and what makes it work?). In M. N. Cooper (Ed.), *Open architecture as communication policy* (pp. 17–40). Stanford, CA: The Center for Internet and Society, Stanford Law School.

Coleman, G. (2014). *Hacker, hoaxer, whistleblower, spy: The many faces of Anonymous.* New York, New York, NY: Verso.

Dant, T. (1999). *Material culture in the social world: Values, activities, lifestyles.* Maidenhead, UK: Open University Press.

Dant, T. (2004). *Critical social theory: Culture, society and critique.* London, UK: Sage.

Deleuze, G. (1992). Postscript on the societies of control. *October, 59*(4), 3–7.

Gehl, R. W. (2014). *Reverse engineering social media: Software, culture, and political economy in new media capitalism.* Philadelphia, PA: Temple University Press.

Gibbons, M., Limoges, C., Nowotny, H., Schwartzman, S., Scott, P., & Trow, M. (1994). *The new production of knowledge: The dynamics of science and research in contemporary societies.* London, UK: Sage.

Grusin, R. (2010). *Premediation: Affect and mediality after 9/11.* New York, NY: Palgrave Macmillan.

Guattari, F. (1989). The three ecologies. *New Formations, 8,* 131–147.

Haggerty, K. D., & Ericson, R. V. (2000). The surveillant assemblage. *The British Journal of Sociology, 51*(4), 605–622.

Herman, A. (2013). Production, consumption, and labor in the social media mode of communication and production. In J. Hunsinger & T. Senft (Eds.), *The social media handbook* (pp. 30–44). London, UK: Routledge.

Hobbes, T. (1994). *Leviathan: With selected variants from the Latin edition of 1668.* Cambridge, MA: Hackett.

Hunsinger, J. (2005). Toward a transdisciplinary internet research. *The Information Society, 21*(4), 277–279.

Hunsinger, J. (2011). Interzoning in after zoning out on infrastructure. *M/C Journal, 14*(5), 7. Retrieved February 25, 2015 from http://journal.mediaculture.org.au/index.php/mcjournal/article/viewArticle/425

Hunsinger, J. (2013). Locus communus: The unconnected in and within virtual worlds. In P. M. A. Baker, J. Hansen, & J. Hunsinger (Eds.), *The unconnected: Social justice, participation, and engagement in the information society* (pp. 125–142). Bern, Switzerland: Peter Lang.

Illich, I. (1977). *Disabling professions.* London, UK: Marion Boyars.

Latour, B. (1993). *We have never been modern.* Cambridge, MA: Harvard University Press.

Luke, T. W. (1995). New world order or neo-world orders: Power, politics and ideology in informationalizing glocalities. In M. Featherstone, S. Lash, & R. Robertsone (Eds.), *Global modernities* (pp. 91–107). London, UK: Sage.

Luke, T. W. (1999). *Capitalism, democracy, and ecology: Departing from Marx.* Champaign: University of Illinois Press.

Nelson, T. H. (1987). *Computer lib/dream machines.* Redmond, WA: Microsoft Press.

Nowotny, H., Scott, P., & Gibbons, M. (2001). *Re-thinking science: Knowledge and the public in an age of uncertainty.* London, UK: Polity Press.

O'Tuathail, G. (1994). (Dis)placing geopolitics: Writing on the maps of global politics. *Environment and Planning D: Society and Space, 12*(5), 525–546.

Raymond, E. S. (2001). *The cathedral & the bazaar: Musings on Linux and Open Source by an accidental revolutionary.* Newton, MA: O'Reilly Media.

Rousseau, J.–J. (1997). *Rousseau: "The Social Contract" and other later political writings.* Cambridge, UK: Cambridge University Press.

Star, S. L. (1999). The ethnography of infrastructure. *American Behavioral Scientist, 43*(3), 377–391.

Star, S. L. (2002). Infrastructure and ethnographic practice: Working on the fringes. *Scandinavian Journal of Information Systems, 14*(2), 107–122.

Virilio, P. (2000). *The information bomb.* New York, NY: Verso.

Virilio, P. (2007). *The original accident.* London, UK: Polity Press.

Virilio, P., & Lotringer, S. (2002). *Crepuscular dawn.* New York, NY: Semiotext(e).

CHAPTER FIVE

Online Performative Identity Theory: A Preliminary Model for Social Media's Impact on Adolescent Identity Formation

Bradley W. Gorham and Jaime R. Riccio

In the first *Produsing Theory* volume, Shayla Thiel-Stern (2012) discussed how adolescents' instant messaging and social media communication routinely breaks the fourth wall to collaboratively produce and perform identity within the confines of both technological and cultural templates. Borrowing from the language of dramaturgy and setting her analysis in symbolic interactionism, she explores how the new ways young people communicate open up new possibilities for presentation of self. We expand that work by adapting and applying social cognitive mechanisms to understand the power and effects of collaborative identity performance and the ways in which repeated and consistent productions of identity create enduring schemas for self. In essence, if symbolic interactionism provides a conceptual way of thinking about interaction online, this chapter aims to link those ideas to the psychological effects such interactions are likely to produce to reinforce perceptions of self around identity.

Children have been of special concern to media effects researchers. Even with the modern conception of media effects as both contingent and subjective (Erbring, Goldenberg, & Miller, 1980; Huang, 2009), American youth are still often viewed as having restricted capacity to resist mediated messages or to be truly active and critical engagers of media content. Although these are valid concerns, new media technologies also grant children access to a variety of tools for defining themselves in ways unique and separate from the adults who worry about them. Children may be at special risk from violent messages on television or the predation of advertisers online (Cai & Zhao,

2010; Comstock & Scharrer, 2007), but they are also open to special rewards offered by new technologies; children's digital media use can positively influence literacy, peer socialization, and self-expression (Anderson-Butcher et al., 2010; Blanchard & Moore, 2010; Davis, 2013).

Young people interact with digital media in four basic ways: consuming content, producing content, sharing content, and communicating with others (Takeuchi, 2011). Social media in particular are relevant for study; because of their affordances, young people's sense of self and identity formation may be more heavily affected by interactive media than by traditional media. Such methods of interaction are well suited for the formation and reification of identity. Media can serve as the "materials out of which we forge our very identities, our sense of selfhood" (Kellner, 1995, p. 5), and indeed, social cognitive theory itself began as an investigation into the ability of media images to influence motivations to enact behaviors in addition to simply learning them. Children spend over half of their days using a variety of media (Brooks-Gun & Donahue, 2008; Lewin, 2010); it follows that children's developing identities are likely influenced not only by *what they see* online but also by the rewards and occasional punishments they perceive for *what they do* online. Social networking sites (SNS), especially, can facilitate the tasks associated with child psychosocial development (Spies Shapiro & Margolin, 2013).

In this chapter, we take preliminary steps toward a theoretical approach to explain why social media play such an important role in how children and adolescents create and perform identity. To adequately understand the psychosocial development of children in the digital age, scholars must better understand the particular affordances (Gibson, 1979) these new technologies allow and how adolescents engage them. With such technologies, children have greater power to determine what media they consume and for what purposes, as well as how they use it to tell stories of themselves. This culture of choice and interactivity has enabled young people to cultivate a variety of unique selves, potentially altering the developmental landscape of youth. We suggest an interactive model, which links a child's relevant personality characteristics, identity performance via social media, peer feedback to and self-evaluation of performance, and the internalization or rejection of identity. We begin with an overview of the key concepts and their relationships.

Online Performative Identity Theory—
An Overview of the Model

Although other scholars have studied the unique way social media influence identity formation and performance (e.g., Cover, 2012), we link sociological understandings of identity and performance with psychological concepts of message processing in a social environment. We understand identities as antecedents: socially constructed sets of performed behaviors that are generally agreed on as consistent with a shared set of beliefs, values, and perspectives. They become meaningful through how people perform and react to them. Most identities already exist in shared cultural spaces, such that children and adolescents come to know what behaviors belong with some identities and not others through their interaction with family, peers, and the media message environment. The message environment includes not only traditional sources of identity-relevant information, such as television and magazines, but also online spaces, such as Facebook, YouTube, Instagram, and so forth.

The message environment, however, does not influence all children equally; personality traits can moderate the relationship between content and effects. Children and teens may be especially open to effects on identity given the nature of adolescence, but this needs to be understood as dependent on personality. Furthermore, this interaction is an ongoing process. Identity construction and reification grow from repeated instances of behavior and feedback. These work to establish which identities become meaningful to the child and which do not; the repeated and consistent nature of these interactions helps build a schema for self.

Therefore, a process develops online in social networking spaces that helps children and adolescents test out and cement identities in the form of self-schemas. Although this is an ongoing, iterative process, for the purposes of building a model, we can think of the *independent variables* as social media use and social media feedback. That is, one key independent variable is online behaviors of self-performance, such as posting status updates, photos, and videos; posting links to stories, music, and videos; and commenting on or liking the posts of others in identity-congruent ways. But equally important in this model is the feedback these self-performative behaviors generate. What other people say about an individual's posts, and their liking or favoriting of posts, is an important variable in this model because this feedback indicates the appropriateness of posts for a particular social identity. That is, feedback helps define which behaviors are congruent with the emerging identity and which are not. In this sense, the technological affordances of the social network allow for a particularly effective system of quickly rendered behavioral rewards to help define a self-schema.

An important *moderating variable* is personality, because not all children or adolescents will orient toward or react to online feedback the same way. Thus, the child's personality plays a key role in influencing the potential impact of social media.

A number of potential *dependent variables* will mark the effects of social media behaviors on identity, including a stronger schema for particular social identities, such that repeated and consistent use leads to a clearer idea of the appropriate behaviors and beliefs for members of a social identity group. Thus, the emerging identity might become more important and more central to a child if identity-congruent online behaviors generate positive feedback, especially if such feedback is relevant and important to the child. Similarly, the importance of a given identity might lessen as expressions of that identity lead to negative (or no) feedback. Further, as an identity becomes more central, children may be more likely to perform identity-congruent behaviors in the future. Thus, we posit that social media use becomes a cycle of identity performance, feedback, and reification as identity-congruent behaviors are rewarded and become more central to the self-schema of the child or adolescent.

With this overview in mind, we now turn to a discussion of the theoretical basis of the conceptual model.

Identity and Performance

One's identity encompasses a sense of continuity and affiliation with or uniqueness from others (Erikson, 1968) and is closely tied to the socialization process and one's social groups (Tajfel & Turner, 1986). According to Baumeister (1988), the importance of social interaction in establishing identity cannot be overlooked, which means the ways in which identities are performed for others are a key part of the socialization process. In the Internet age, and especially for young people, social groups are increasingly found online. Because of the interactive and performative nature of online social interaction, we can examine identity formation from psychological, ecological, symbolic-interactionist, and dramaturgic perspectives.

In psychobiological approaches to development, a child's temperament, emotion, and self are key factors in identity development (Rothbart & Ahadi, 1994). Rubin, Coplan, Chen, Bowker, and McDonald (2010) noted that sense of self emerges relatively early in a child's life but is shaped and reshaped throughout. The central component to the construct of self is one's sense of self-awareness. Self also includes self-representation (who one thinks one is), self-evaluations based on others' evaluations, and the social self (regarded in a social context with concern for the assessment of others). The latter factors are of key importance to young adolescents.

The environmental context in which a young person exists greatly influences the processes of development (Bronfenbrenner, 1995). A child's development requires that particular experiences occur during "relevant sensitive periods" in his or her life (Rutter & O'Connor, 2004, p. 84). Sensory and behavioral patterns emerge as the brain learns from experiences.

Today, these experiences are often socially mediated. Because children become most sensitive to peer relationships (as opposed to relationships with parents) during preadolescence (Harris, 1995), young people's social media use becomes key. Peer interaction and feedback, such as occurs on SNS, can become the most powerful predictors of identity (Davis, 2013).

The influence of online peer relationships can be considered a form of informal social control and cohesion. Leventhal and Brooks-Gunn (2000) called this "collective efficacy," defined as social connections arising from mutual trust and shared values. The collective efficacy present in children's interactions on social media may assist in the performance, internalization, or rejection of various identities, which may then influence the cementing of a particular identity or the experimenting with new ones. Consistent with a symbolic interactionist tradition (Mead, 1934)—which assumes that human behavior is based on meaning created during social interaction and posits that meaning making is a behavioral process rather than a cognitive one—a child's interactions with mediated others today has become a necessary step in the process of identity development. In fact, Mead himself found that children's abilities to organize and reflect on their identities develop as a function of peer interaction (Rubin et al., 2010).

Although identity may be thought of as an internal state of belief (i.e., how one sees oneself), it must also be understood as a *performance,* a presentation of self for others to see and interact with. Self-presentation has been theorized in many settings, originating in sociology with Goffman's impression management, which links symbolic interactionism with dramaturgy, or the analysis of our public presentations of self. According to Goffman (1959), a person's sense of self is formed through his or her interactions with others. This interaction is part of a performance metaphor, in which the actor constructs a role for a receptive audience. This "everyday acting" (p. 24) occurs in the management of one's impression (or the identity that he or she presents to others). The concept of impression management outlines how one attempts to control the meanings others create about him or her during these performative interactions, transforming a person into a number of roles like an actor on a stage. Borrowing the theater metaphor, a dramaturgical approach posits that all human interaction falls on a spectrum of performance (Ritzer, 2007), with individuals moderating their public, "on stage" interactions to produce the desired impressions. Goffman perceived a world in which dramatic interactions with "like-minded others" form and strengthen consensus (Edgley, 2003; Goffman, 1959, p. 42), including the meaning of or belief in one's own identity. In this way, one's identity is something formed through performance, shifting based on one's self-presentation.

Similarly, Judith Butler (1990) theorized gender as socially contrived performance rather than socially constructed identity, popularizing a notion of performativity as the essential nature of identity. To Butler, identity is not inherent in a person, existing and giving rise to behaviors, but instead is a *product* of the socially sanctioned performances in which a person engages. As Cover (2012) noted, performativity is "identity produced through the citation of culturally given identity categories or norms in a reiterative process" (p. 180). Although we do not believe we must abandon the notion of identity as a cognitive sense of self, Butler reminds us that it can be an effect more than a cause. This seems especially relevant to a discussion of adolescent identity construction and the role of social media.

Children and adolescents use narratives to structure their experiences, telling stories about their daily lives, their aspirations, and who they want to be (Ahn, 2011). The profile pages and status updates via social media are tools allowing children to craft various narratives about themselves and their world. In so doing, they define and then share these personas.

The Affordances of Social Media in Identity Formation

Technological affordance refers to the types of actions technologies allow and the possible ways in which these technologies will be utilized by their users, including actions that are novel or unrelated to the technology's intended use (Majchrzak, Faraj, Kane, & Azad, 2013). The affordances of social media make them an ideal performance space due to the various dramaturgical props available: profile images, photos, videos, links to other content, status updates, check-ins, likes, comments, and more. No space in modern time seems more fitting for the presentation and performance of self than the zone of synergistic production and reception presented by SNS. Children can use the tools to present themselves as they wish in order to perform an ideal self (Goffman, 1959) and showcase these preferences and performances via the Internet. Such constant performance and presentation of self is now incorporated into a child's socialization, and when a social media platform is suited to a child's cognitive stage, it can promote the development of a number of capacities, including those of identity formation (Bittman, Rutherford, Brown, & Unsworth, 2011; Heintz & Wartella, 2012).

The significance of these sites, therefore, is not simply their extreme popularity among young people but rather how the technological affordances they offer their young users facilitate a form of social grooming. Social grooming (Dunbar, 1998) refers to the process of establishing and maintaining relationships and relational hierarchies through visible interaction. Online social media platforms afford many ways in which users can engage in social grooming behaviors to maintain and reinforce social relationships (Utz & Beukeboom, 2011). By commenting on other users' photos, liking their status updates, or posting links to the walls of others, young people can effectively groom the social relationships that matter to them and present themselves in what is deemed appropriate relational terms. Such online grooming behaviors have already been found to have effects. In a study of users' behaviors on Facebook, Kim and Chock (2014) found that these behaviors can influence self-concept: Increased exposure to others' profiles during social grooming interactions affected the body images of their young subjects

Because SNS allow their users a broad range of actions for self-expression, they also afford a great many opportunities for others to provide feedback on those expressions. Feedback is an integral part of the process of meaning making in these spaces, as others comment on and react to one's posts and expressions. Therefore, SNS also afford the ability to police what kinds of expressions and posts are deemed appropriate, desirable, or problematic. These technological spaces afford a system of yays and nays, of rewards and punishments, thereby drawing the boundaries for appropriate expressions of self. This has clear implications for identity formation, because it ties the use of these online spaces to larger forces of social cognition.

Children's Social Media Use and Identity

Bandura's (2009) social cognitive theory of communication argued that communication and communication effects must be understood within their social contexts. The mere modeling of behaviors in the communication sphere is not enough to assume that (if the context allows) behaviors will be learned, internalized, and reproduced; instead, behaviors must be situated in a larger context of motivation. Are behaviors rewarded? Are they punished? When performed by people perceived as "like me," are such actions praised or criticized by relevant peers? Is there an incentive to reproduce behaviors should the context allow? Communication is thus situated in the social space of communicators, and the visible reactions of peers become a potentially powerful influence on the reproductions of communicative action. Importantly, the individual plays a role in choosing and creating these spaces, so the selections people make in choosing some communication environments over others help determine the range of messages they encounter and the contexts of motivations.

The relationship between the social context of communication and the understanding of identity has been particularly well theorized in relation to gender, revealing interactions among biological sex, socially constructed gender, and gender as performance. Bussey and Bandura (1999), for example, suggested in their social cognitive theory of gender that the selection of gendered media plays an important role in developing gendered self-concepts. Societal norms, presented through repeated and consistent media examples, help develop relatively stable gender schemas (Bem, 1981) about what is considered masculine and feminine. Even if biological sex plays an important role in pushing individuals toward some types of media content rather than others, there is still a role for an individual's selective exposure to different forms of gender expression in media—but those choices are nonetheless likely to be strongly influenced by an individual's acceptance of social norms. For example, in their study of the self-selection of gendered magazines, Knobloch-Westerwick and Hoplamazian (2012) found that both biological sex and gendered self-concept influenced magazine selection. Overall, males preferred to browse and read masculine-stereotyped magazines more than women, but even among males, men with more masculine self-concepts were more likely to select masculine-themed magazines compared to men with less masculine self-concepts.

But although selectivity influences the communication environment, much research demonstrates the constraining power of social norms and expectations. The gendered nature of the communication environment means that individuals receive strong messages about gender roles and their appropriateness (Eagly, Beall, & Sternberg, 2004); if gender encourages us to see some messages as more relevant for the self than others, we should hardly be surprised that self-schemas (Markus, 1977)—the mental frameworks we carry about ourselves—should reflect and help perpetuate a gendered communication environment (Markus, Crane, Bernstein, & Siladi,

1982), both online and in the physical world. Knobloch-Westerwick and Hoplamazian (2012) argued for an interactive relationship among gender, identity, selectivity, and gender conformity that can be usefully expanded here. Self-identity schema at the cognitive level refers to the intersection of traits, actions, and roles that we believe are appropriate for us. They develop through repeated interactions with others and with media content that tells us, with varying degrees of consistency, what certain identities mean and how well we conform to the expectations associated with those identities. The degree to which we perceive identity-related actions and performances—our own and others'—as rewarded or punished by the media environment around us influences the perceived attractiveness of those performances and therefore the motivation to reproduce those performances.

Personality and Social Media

What social cognitive theory reminds us, then, is that although the nature of media content plays an important role in shaping the contours of media experience, so does how that content is perceived. How one interprets the reactions of one's peers online, for instance, is necessarily moderated by one's personality, and social media use has been highly correlated with personality traits (Amichai-Hamburger & Vinitzky, 2010; Ryan & Xenos, 2011). Because of the ability to display oneself to the world, frequent SNS users tend to rank highly in personality traits such as extraversion and narcissism. Evidence of social media's ability to cultivate such characteristics (Correa, Hinsley, & de Zúñiga, 2010) points to the possible recursive relationship between personality traits and social media use.

Thus, children come to social media interactions with preexisting personality traits that may influence why and how they use SNS. John and Srivastava (1999) presented a model of prominent personality traits that have been linked to social media use. Their Five Factor Model posits themes covering a broad range of personality characteristics, including openness to experience, conscientiousness, extraversion, agreeableness, and neuroticism. *Openness to experience* can be defined as one's emotionality, curiosity, and imagination. It is sometimes also thought of as a level of intellect or independence, with those ranking high in curiosity or information seeking on one end of the spectrum and people who are more cautious placed at the other (McCrae & John, 1992). Individuals who have greater openness to experience have been shown to use SNS more frequently and to use a greater variety of features offered by such sites (Amichai-Hamburger & Vinitzky, 2010; Correa et al., 2010).

Conscientiousness refers to one's orderliness, work ethic, or preference for planning versus the ability to be spontaneous or easygoing. Conscientious individuals have been found to be more avoidant of social media, considering it a form of distraction (Butt & Phillips, 2008; Ryan & Xenos, 2011).

Extraversion denotes one's assertiveness and sociability or lack thereof (Goldberg, 1992). SNS users tend to rank high in extraversion and to have more online friends (Amichai-Hamburger & Vinitzky, 2010; Ross et al., 2009).

Agreeableness is the tendency for one to be trustworthy or compassionate versus detached or uncooperative. Research has not yet found a link between social media use and agreeableness, although greater numbers of friends may be associated with agreeableness (Ross et al., 2009).

Neuroticism is high vulnerability or anxiety, as opposed to being emotionally stable or confident (Digman, 1990). Research has shown that those high in neuroticism tend to be frequent users of Internet technologies and social media (Butt & Phillips, 2008; Ryan & Xenos, 2011).

Research into these five factors of personality posits that social media use may be part of a cycle fueled by personality characteristics that influences identity formation, affecting further personality development. In a model of children's social media use, personality traits may be a significant driver of online interactions in a system of social media use and identity development that is simultaneously constrained by that system's social feedback mechanisms.

Online Identity Development

In the digital world, children are social actors with, perhaps, more individual agency than in the physical world. They take an active role in their identity development. Digital technologies aid this role as structures not only shaping children's development but also allowing children to shape their own. Socialization is key for successful identity development, and sharing meaning among peers in a socially mediated setting allows young people to test identities and performances on willing and eager audiences of fellow experimenters. The Internet has been called an "unrestricted laboratory for identity experimentation" (Williams & Merten, 2008, p. 256). For children, the Internet is perhaps their most unrestricted world: one with limited interference from the rule systems of parents, schools, and other outside forces. As Kjartan Ólafsson (2011) argued, "although the internet is central to current studies of children's media use, it must be understood as part of a bigger picture" (p. 365).

Adolescents are especially sensitive to their surroundings and aim to match their outer selves with their inner meanings of self (Gross, 1987). Erikson (1968) called this a period of *fidelity,* in which the identity is being forged and a child attempts to cement his or her role in society. If children are reluctant to commit to any particular role or identity being presented, "identity crisis" or "role confusion" may result (Stevens, 1983, p. 49). The affordances of SNS may help children not only perform a role that later becomes an identity but also clarify any confusion that results, with the aid of constant and immediate feedback for an age group that values it the most.

Besides influencing the fidelity of identity, the dynamics of online self-presentation can lead to what is called "identity shift," or the modification of a user's personality following the performance of an identity (Walther et al., 2011). Identity shift is influenced both by the act of performing an identity over time and by the feedback from other SNS users. The audience may or may not respond to posts, thus reaffirming or disavowing a given performance. Gonzales and Hancock (2008) found that selective self-presentation online may lead to identity internalization and transformation. Walther et al. (2011) argued that this is largely due to the public nature of self-performance on SNS: "When one is publically identifiable, one's self-presentation links the characteristics that one performs to one's identity" (p. 6). In other words, the social cognitive mechanism of evaluating the outcomes of given behaviors to determine the motivation to reproduce those behaviors is at significant play in the socially mediated world and can thus greatly influence the evolving nature of identity. As more online behaviors are performed and responded to, a schema for self can begin to emerge, as patterns of call and response arise from these online interactions.

Online Performative Identity Theory—Next Steps

What we are calling online performative identity theory is thus a framework for thinking about how new communication technologies play an important role in the formation, construction, and reification of identity among young people. It is online in the sense that these particular behaviors that are so important for the formation of identity can really only take place in the mediated spaces of social media, with their asynchronous ability to allow many people from multiple social spaces to contribute to a cycle of communication. It is performative because it relies not on the observation of behaviors (as in traditional media) but on enacting and practicing behaviors in a conscious way for others to see. It is an identity theory because it argues that the sense of self is greatly influenced by these interactions, and thus the possibilities and constraints of the communication environment and the selectivity of the performers in that environment ultimately shape how individuals see and understand themselves.

This preliminary presentation of online performative identity theory suggests a number of directions for further exploration. For one, the mechanisms by which this process unfolds can be further explicated and examined. What are the relationships between specific personality traits and specific social media behaviors? Are more dramaturgical online behaviors consistent with only certain kinds of personalities? As new platforms emerge with different affordances, such as the ephemeralness of content on Snapchat or the anonymity of Yik Yak, do they attract certain kinds of personalities over others?

Just as personality traits are expected to have different relationships with these social tools, another fruitful avenue of inquiry involves the identities themselves. Do identities around socially significant categories of race, ethnicity, gender, or sexual identity develop differently through social media performance compared to identities around lifestyle choices, hobbies, artists, or sport fandom? Is it easier for adolescents to identify with majority identities, and to what extent does the size and composition of one's social network affect the identity formation process? The online performative identity theory framework suggests ways in which all of these questions can be explicated and studied.

References

Ahn, J. (2011). Review of children's identity construction via narratives. *Creative Education, 2*(5), 415–417.

Amichai-Hamburger, Y., & Vinitzky, G. (2010). Social network use and personality. *Computers in Human Behavior, 26,* 1289–1295.

Anderson-Butcher, D., Lasseigne, A., Ball, A., Brzozowski, M., Lehnert, M., & McCormick, B. (2010). Adolescent weblog use: Risky or protective? *Child & Adolescent Social Work Journal, 27*(1), 63–77.

Bandura, A. (2009). Social cognitive theory of mass communication. In J. Bryant & M. B. Oliver (eds.), *Media effects: Advances in theory and research* (3rd ed.). New York, NY: Routledge.

Baumeister, R. F. (1998). The self. In D. T. Gilbert, S. T. Fiske, & G. Lindzay (Eds.), *The handbook of social psychology* (4th ed., Vol. 2, pp. 680–740). Boston, MA: McGraw-Hill.

Bem, S. L. (1981). Gender schema theory: A cognitive account of sex typing. *Psychological Review, 88,* 354–364.

Bittman, M., Rutherford, L., Brown, J., & Unsworth, L. (2011). Digital natives? New and old media and children's outcomes. *Australian Journal of Education (ACER Press), 55*(2), 161–175.

Blanchard, J., & Moore, T. (2010). *The digital world of young children: Impact on emergent literacy.* Retrieved April 25, 2015 from http://www.pearsonfoundation.org/downloads/ EmergentLiteracy-WhitePaper.pdf

Bronfenbrenner, U. (1995). The bioecological model from a life course perspective: Reflections of a participant observer. In P. Moen, G. H. Elder, Jr., & K. Luscher (Eds.), *Examining lives in context: Perspectives on the ecology of human development* (pp. 599–618). Washington, DC: American Psychological Association.

Brooks-Gun, J., & Donahue, E. H. (2008). Introducing the issue. *The Future of Children, 18*(1), 3–10.

Bussey, K., & Bandura, A. (1999). Social cognitive theory of gender development and differentiation. *Psychological Review, 106,* 676–714.

Butler, J. (1990). *Gender trouble: Feminism and the subversion of identity.* London, UK: Routledge.

Butt, S., & Phillips, J. G. (2008). Personality and self reported mobile phone use. *Computers in Human Behavior, 24*(2), 346–360.

Cai, X., & Zhao, X. (2010). Click here, kids! Online advertising practices on popular children's websites. *Journal of Children and Media, 4*(2), 135–154.

Comstock, G., & Scharrer, E. (2007). *Media and the American child.* Philadelphia, PA: Academic Press.

Correa, T., Hinsley, A., & de Zúñiga, H. (2010). Who interacts on the Web?: The intersection of users' personality and social media use. *Computers in Human Behavior, 26*(*2*), 247–253. doi:10.1016/j.chb.2009.09.003

Cover, R. (2012). Performing and undoing identity online: Social networking, identity theories and the incompatibility of online profiles and friendship regimes. *Convergence: The International Journal of Research into New Media Technologies, 18*(2), 177–193.

Davis, K. (2013). Young people's digital lives: The impact of interpersonal relationships and digital media use on adolescents' sense of identity. *Computers in Human Behavior, 29*(6), 2281–2293.

Digman, J. M. (1990). Personality structure: Emergence of the five-factor model. *Annual Review of Psychology, 41,* 417–440.

Dunbar, R. (1998). *Grooming, gossip, and the evolution of language.* Cambridge, MA: Harvard University Press.

Eagly, A. H., Beall, A. E., & Sternberg, R. J. (2004). *The psychology of gender.* New York, NY: Guilford Press.

Edgley, C. (2003). The dramaturgical genre. In L. Reynolds & N. Herman-Kinney (Eds.), *The handbook of symbolic interactionism* (pp. 141–172). Walnut Creek, CA: AltaMira Press.

Erbring, L., Goldenberg, E. N., & Miller, A. H. (1980). Front-page news and real-world cues: A new look at agenda-setting by the media. *American Journal of Political Science, 24,* 16–49.

Erikson, E. H. (1968) *Identity: Youth and crisis.* New York, NY: W. W. Norton.

Gibson, J. (1979). *The ecological approach to visual perception.* Boston, MA: Houghton

Mifflin.

Goffman, E. (1959). *The presentation of self in everyday life.* New York, NY: Doubleday.

Goldberg, L. R. (1992). The development of markers for the big-five factor structure. *Psychological Assessment, 4,* 26–42.

Gonzales, A. L., & Hancock, J. T. (2008). Identity shift in computer-mediated environments. *Media Psychology, 11,* 167–185.

Gross, F. L. (1987). *Introducing Erik Erikson: An invitation to his thinking.* Lanham, MD: University Press of America.

Harris, J. R. (1995). Where is the child's environment? A group socialization theory of development. *Developmental Review, 102*(3), 458–489.

Heintz, K. E., & Wartella, E. A. (2012). Young children's learning from screen media. *Communication Research Trends, 31*(3), 22–29.

Huang, C. (2010). Internet use and psychological well-being: A meta-analysis. *Cyberpsychology, Behavior, and Social Networking, 13*(3), 241–249.

John, O. P., & Srivastava, S. (1999). The Big-Five trait taxonomy: History, measurement, and theoretical perspectives. In L. A. Pervin & O. P. John (Eds.), *Handbook of personality: Theory and research* (Vol. 2, pp. 102–138). New York, NY: Guilford Press.

Kellner, D. (1995). *Media culture: Cultural studies, identity and politics between the modern and the postmodern.* London, UK: Routledge.

Kim, J. W., & Chock, M. (2014, November). *Facebook and body image: Relationships between social grooming in social media and the drives for thinness and muscularity.* Paper presented at the annual conference of the National Communication Association, Seattle, WA.

Knobloch-Westerwick, S., & Hoplamazian, G. J. (2012). Gendering the self: Selective magazine reading and the reinforcement of gender conformity. *Communication Research, 39*(3), 358–384.

Leventhal, T. & Brooks-Gunn, J. (2000). The neighborhoods they live in: The effects of neighborhood residence on child and adolescent outcomes. *Psychological Bulletin, 126*(2), 309–337.

Lewin, T. (2010, January 20). If your kids are awake, they're probably online. *The New York Times.* Retrieved February 27, 2015 from http://www.nytimes.com/2010/01/20/education/20wired.html?_r=0

Majchrzak, A., Faraj, S., Kane, G. C., & Azad, B. (2013). The contradictory influence of social media affordances on online communal knowledge sharing. *Journal of Computer-Mediated Communication, 19,* 38–55.

Markus, H. (1977). Self-schemata and processing information about the self. *Journal of Personality and Social Psychology, 35*(2), 63–78.

Markus, H., Crane, M., Bernstein, S., & Siladi, M. (1982). Self-schemas and gender. *Journal of Personality and Social Psychology, 42*(1), 38–50.

McCrae, R. R., & John, O. P. (1992). An introduction to the five-factor model and its applications. *Journal of Personality, 60*(2), 175–215.

Mead, G. H. (1934). *Mind, self and society: From the standpoint of a social behaviorist.* Chicago, IL: The University of Chicago Press.

Ólafsson, K. (2011). Is more research really needed? Lessons from the study of children's internet use in Europe. *International Journal of Media & Cultural Politics, 7*(3), 363–369.

Ritzer, G. (2007). *Contemporary sociological theory and its classical roots: The basics.* New York, NY: McGraw-Hill.

Ross, C., Orr, E. S., Sisic, M., Arseneault, J. M., Simmering, M. G., & Orr, R. R. (2009). Personality and motivations associated with Facebook use. *Computers in Human Behavior, 25*(2), 578–586.

Rothbart, M. K., & Ahadi, S. A. (1994). Temperament and the development of personality. *Journal of Abnormal Psychology, 103,* 55–66.

Rubin, K., Coplan, R., Chen, X., Bowker, J., & McDonald, K. L. (2010.) Peer relationships in childhood. In M. H. Bornstein & M. E. Lamb, (Eds.), *Developmental science: An advanced textbook* (6th ed., pp. 519–570). Mahwah, NJ: Lawrence Erlbaum.

Rutter, M., & O'Connor, T. G. (2004). Are there biological programming effects for psychological development? Findings from a study of Romanian adoptees. *Developmental Psychology, 40*(1), 81–94.

Ryan, T., & Xenos, S. (2011). Who uses Facebook? An investigation into the relationship between the Big Five, shyness, narcissism, loneliness, and Facebook usage. *Computers in Human Behavior, 8,* 1658–1664.

Spies Shapiro, L. A., & Margolin, G. (2013). Growing up wired: Social networking sites and adolescent psychosocial development. *Clinical Child & Family Psychology Review, 17,* 1–18.

Stevens, R. (1983). *Erik Erikson: An introduction.* New York, NY: St. Martin's Press.

Tajfel, H., & Turner, J. C. (1986). The social identity theory of intergroup behaviour. In S. Worchel & W. G. Austin (Eds.), *Psychology of intergroup relations* (pp. 7–24). Chicago, IL: Nelson-Hall.

Takeuchi, L. (2011). *Families matter: Designing media for a digital age.* Retrieved February 27, 2015 from http://www.joanganzcooneycenter.org/wp-content/uploads/2011/06/jgcc_families matter.pdf

Thiel-Stern, S. (2012). Collaborative, productive, performative, template: Youth, identity, and breaking the fourth wall online. In R. A. Lind (Ed.), *Produsing theory in a digital world: The intersection of audiences and production in contemporary theory* (pp. 87–103). New York, NY: Peter Lang.

Utz, S., & Beukeboom, C. J. (2011). Grooming, gossip, Facebook and Myspace. *Information, Communication and Society, 11*(4), 544–564.

Walther, J. B., Liang, Y., Deandrea, D. C., Tong, S. T., Carr, C. T., Spottswood, E. L., & Amichai-Hamburger, Y. (2011). The effect of feedback on identity shift in computer-mediated communication. *Media Psychology, 14*(1), 1–26.

Williams, A. L., & Merten, M. J. (2008). A review of online social networking profiles by adolescents: Implications for future research and intervention. *Adolescence, 43*(170), 253–274.

Understanding the Popularity of Social Media: Flow Theory, Optimal Experience, and Public Media Engagement

John V. Pavlik

S ocial media are a wildly popular form of communication. Tweeting, posting to Facebook, and sharing photos on Instagram or Pinterest and other forms of social media engage 73% of the U.S. public, according to data from the Pew Research Internet Project (Lunden, 2013). Social networking media are also popular around the world. For instance, estimates are that at least half of China's 600 million or so Internet users utilize social media (Mei, 2012). Beyond China, social media are also increasingly popular in many other regions. In Qatar and elsewhere in the Gulf region of the Middle East, North Africa and beyond, social media, especially Facebook, Twitter, and Instagram, are heavily used across many social strata (Northwestern University in Qatar, 2014). Globally, data suggest at least one in four persons uses social media around the world (eMarketer, 2013). In 2014, Yahoo estimates people will take 880 billion photos, many of them selfies posted on social media (Agence France Press, 2013).

With the dramatic growth of social networking media, social scientists have sought explanations for the adoption and appeal of this seemingly new form of social engagement. Some point to the potential capability of using social media to engage in political activism. Others suggest that even more fundamentally, humans are naturally social creatures, leading them to use social media to obtain certain gratifications (Ancu & Cozma, 2009; Ballard, 2011; Ellison, Steinfield, & Lampe, 2007; Gallion, 2010; Urista, Dong, & Day, 2008). It is therefore logical, they often suggest, that people would be drawn to new tools that enable them to communicate more easily, efficiently, and rapidly with their friends and family. This uses and gratifications approach may provide some understanding of the motivations for social media use, but it leaves room for additional theoretical analysis of their extraordinary growth and popularity.

Flow Theory

An alternative perspective comes in the form of the psychological theory of flow. In contrast to much psychological theory that has focused on what hinders health and quality of life, flow theory offers a positive psychological framework (Seligman & Csikszentmihalyi, 2000). Psychologist Mihaly Csikszentmihalyi (1990) introduced the theory of flow about the same time the World Wide Web made its public debut in 1990 (World Wide Web Foundation, 2008). Flow theory provides insight into the psychology of optimal human experience. It provides a compelling theoretical foundation for understanding *why* people like to use social networking media. It is completely unrelated to the theory of flow TV as proposed by media studies scholar Raymond Williams (1974) and others, which critiques the commercial nature of television as programmers seek to hold the audience from one program or segment to the next, especially in a U.S. context.

As Csikszentmihalyi (1997) proposed it, flow refers to a positive, pur-pose-driven mental state in which an individual experiences a feeling of being fully immersed, highly focused, and absorbed in an activity or task, deriving enjoyment from the process of the activity itself. Flow is about directing, and learning to direct and focus attention. This is an essential ingredient in the individual's social media experience, as persons young and old often become highly focused on using social media to communicate with family and friends throughout the day.

Attention is a fundamental concept in understanding flow. For our pur-poses, we use the classic definition of attention as articulated by psychologist William James (1890). Attention "is the taking possession by the mind, in clear and vivid form, of one out of what seem several simultaneously possible objects or trains of thought. Focalization, concentration, of con-sciousness are of its essence. It implies withdrawal from some things in order to deal effectively with others, and is a condition which has a real opposite in the confused, dazed, scatterbrained state which in French is called distrac-tion, and Zerstreutheit [absentmindedness] in German" (p. 403).

Flow is about focusing attention, organizing, or helping to increase order in one's consciousness. Yet flow is also about being in tune with one's physical surroundings and developing a feeling of being part of a larger whole. Csikszentmihalyi (1990) defined flow as "an intrinsically motivated, task-focused state characterized by full concentration, a change in the awareness of time (e.g., time passing quickly), feelings of clarity and control, a merging of action and awareness, and a lack of self-consciousness" (p. 433). Two hallmarks of the flow experience, particularly deep or intense flow, are (a) a feeling of spontaneous joy or possibly rapture and (b) an altered sense of time (Goleman, 1996).

In some flow experiences involving an activity, such as an athletic competition, time seems to slow, and this could occur for persons actively engaged in the use of social media. An athlete in a flow state seems to see the field more clearly and the pace of gameplay seems to slow. In other cases, an activity that might last for hours seems to pass in just a few minutes for a person in an intense or deep flow state. I myself experienced such a deep flow state. While scuba diving in the water-filled cavern of the Dos Ojos cenote in Mexico's Yucatán Peninsula, exploring its many otherworldly features, including an ancient Mayan altar, I felt an almost oneness with the water and surroundings. My family reported similar feelings. We swam and dove for what seemed to be about an hour. Yet, when we emerged and climbed back into the neighboring jungle, we were surprised to find more than five hours had passed. Intense flow psychological states can have an almost transcendental quality for the individual as she or he can feel a oneness or close harmony with her or his surroundings. In this manner, people can almost lose track of their own selves during an intense flow state. During intense flow, thinking can be substantially clearer, more insightful, and focused; the individual experiences a feeling of calmness, tranquility, and relaxation, yet heightened productivity and quality.

In an athletics context, Kotler (2014) outlined the role flow plays in advancing human performance to even faster, farther, and stronger achievement. Kotler noted how, in a wide range of athletics, from professional sports to the Olympics, many athletes are turning to psychological methods to enhance their ability to achieve the flow state and thereby elevate their performance. This is especially the case for older athletes who may no longer be at their physical peak, but by concentrating their focus, they can achieve a more optimal experience. Kotler's evidence reflects an increasing recognition that flow may explain much in human behavior (Schaffer, 2013).

As such, flow has potential application to a wide variety of fields, including not only sports and athletic endeavors but also the arts, writing, education, gameplay, music, religion, spirituality, and virtually any work performance (Chen, 2008; Nakamura & Csikszentmihalyi, 2001; Sherry,

2006). In conducting international fieldwork on flow, Csikszentmihalyi found that people around the world often experience flow across a wide variety of tasks, ranging from rock climbing to musical performance, writing poetry to playing chess. Larson (1988) reported evidence linking flow to writing.

Flow also can pertain to human relationships, including those in a digital or social media network and other media use (Kubey & Csikszentmihalyi, 1990). Flow at least partially explains the compelling attractiveness, almost compulsiveness, and sometimes even addiction of watching television, at least for some. Television watching can serve as a means to provide order (albeit at times at a low level) to one's consciousness. It requires little effort on the part of the viewer, other than to watch or listen to what is being presented on the screen. Flow can occur, especially during sports viewing, but deep flow is unlikely, because there is generally little engagement of the viewer in the task of watching TV. Research suggests that because viewing television is largely a passive task, it is more apt to lead to apathy (1990) than to flow. In contrast, social media represent a more active form of human communication, and as such, they may bring the potential for a much more intense flow experience for the user.

Flow states may be achieved alone or with others when engaged in collaborative activities that meet the conditions described earlier. Csikszentmihalyi (1997) found that several qualities can encourage flow experiences among group activities. Among the qualities are working in parallel, organized activities, prototyping, and visualization. This work has particular relevance to teaching, collaborative work, and social media.

Research also shows that flow states do not involve drug or chemical use, although research also suggests that drug use is likely to inhibit achieving a flow state or its benefits for human experience (Nieoullon, 2002). However, it is possible that achieving a flow state may trigger chemical reactions within the body, much as certain athletic activity can trigger the release of endorphins or dopamine (Stewart, 2000). Research has not yet documented such a chemical effect of flow states, but it does raise the question of whether neurochemical processes may be involved (Marr, 2001).

Chen, Wigand, and Nilan (1999) documented the role of flow theory in users' certain online activities. Their research verified the experience of flow on the web: "Web users' flow experiences are multi-dimensional and flow is a complicated construct. Any inquiries into this experience should not limit examination to a single or only a few dimensions" (p. 279). Other research relates flow experience with web and other online activity (Novak & Hoffman, 1997; Novak, Hoffman, & Yung, 1998). Some earlier research was not able to delineate the existence of flow experiences while using computers more generally (Ghani & Deshpande, 1994; Ghani, Supnick, & Rooney, 1991; Webster, Trevino, & Ryan, 1993). Meanwhile, Kraut et al. (1998) found that greater use of the Internet is linked to greater depression and loneliness. This research also suggests Internet use is associated with decreased social involvement and psychological well-being.

It is also important to consider the concept of flow within an ethics context. It is certainly possible that an individual with high skills and challenges to match may be directing her or his energy toward an unethical or even illegal purpose. Moreover, the flow state can become an addiction. This can pertain to playing video games or other active media pursuits and some binge viewing of television (Kubey & Csikszentmihalyi, 2002). The potential negative consequences of flow in a media context, or the possible socially harmful consequences of flow and social media, are not insignificant. Kubey and Csikszentmihalyi documented the sometimes addictive nature of television viewing. Tannenbaum (1980) wrote the following long before social media had even been invented, yet his words have a particularly familiar ring in a social media context: "Among life's more embarrassing moments have been countless occasions when I am engaged in conversation in a room while a TV set is on, and I cannot for the life of me stop from periodically glancing over to the screen. This occurs not only during dull conversations but during reasonably interesting ones just as well" (p. 135).

Research shows that "obsessive passion is either unrelated to or negatively associated with indicators of well-being during, after, and when prevented from engaging in the activity" (Carpentier, Mageau, & Vallerand, 2012, p. 502). Other research tends to confirm this (Mageau & Vallerand, 2007; Philippe, Vallerand, & Lavigne, 2009; Rousseau & Vallerand, 2003, 2008; Vallerand et al., 2007, 2008). This suggests that if an individual becomes obsessed with social media, the behavior might very well become harmful to personal well-being. As my 20-year-old daughter pointed out, social media can sometimes even intervene to disrupt the individual's flow state. Texts and other social media can intrude and draw someone out of living in the moment. Beeping mobile devices and flashing text messages might sometimes serve more to inhibit the flow than encourage it.

Seeking Order: Conditions Shaping Flow

Flow theory posits that humans have a strong inclination to seek order in their consciousness. Research suggests that among the most powerful means to increase that order is to enter into goal-directed behavior under certain conditions (Rathunde & Csikszetnmihalyi, 2005): (a) variety in the activity, (b) clear goals for the individual, (c) flexible and appropriate challenges, (d) immediate feedback, (e) an alignment of the challenges required and the individual's level of skill, (f) an autotelic personality, and (g) a feeling of being connected to a larger whole.

Csikszetnmihalyi (1997) presented a model of the alignment challenges and skills needed to achieve the flow state and the consequences of flow. This model assumes the first four flow conditions are met, including variety, clear goals, flexibility, and immediate feedback. When an alignment of skill and challenge occurs, the individual can achieve a flow state. Flow can be even further enhanced if the sixth condition is met, as is discussed later. While in a flow state, the individual's psychic energy is positively directed and not wasted by distractions, frustrations, or even anomie (Durkheim, 1893). Consequently, the individual associates positive feelings with the activity in which flow is achieved and will seek to engage in that behavior more often or for longer periods of time. Social networking media are remarkably attuned to these conditions.

Csikszentmihalyi's model is ordered along two axes, challenge level and skill level, ranging from low to high. It suggests that there are multiple psychological states associated with the flow condition. Flow, especially deep flow, can result when the conditions outlined earlier are maximized, challenges and skills continue to grow in parallel fashion, and high levels of each are maintained. Conversely, under conditions of low skill and challenge, such as watching television, the individual is likely to experience apathy. Moderate skill aligned with low or moderate challenge may lead to boredom, although high skill and low or moderate challenge may produce a feeling of relaxation. In this condition, the individual enjoys the ease of performing a task but does not achieve a deep flow state. High skill and moderate challenge may also produce a feeling of control for the individual. Meanwhile, in conditions where the challenge may exceed the user's skill, other more negative feelings are likely to emerge. Where skills are low but the challenge is moderate, the individual may feel worried or even anxious if the challenge grows even greater in relation to the individual's skill level. When the challenge is high and the individual's skill level is moderate, arousal may result, providing a strong motivation to increase one's skill level.

Research suggests a seventh condition is sometimes especially relevant to achieving a deep flow state (Csikszentmihalyi, 1990). This condition occurs when the activity allows the individual to feel connected to a larger whole. For instance, athletic competition can facilitate the flow state by enabling participants and even fans to feel connected not only to their team but also to a larger community, as evident during the 2014 FIFA World Cup football championship played in Brazil (Mirror, 2014).

Variety

The variety in social media is nearly limitless, because users can compose an almost infinite range of text messages, shoot photos and video of almost anything, and post and repost content of all types on their mobile or social media network. Through network-connected mobile devices, people can engage in social media virtually anytime and anywhere. Consider as an illustration the phenomenal popularity of the ice bucket challenge that emerged in the summer of 2014. In the challenge, each person video records her or his unique means of implementing the challenge and then shares the video on YouTube. CBS News reported that as of August 20, 2014, 2.4 million people had accepted the challenge, with some 28 million following or viewing their challenges on social media. In the process, these ice bucket challenge participants generated some $94 million for ALS from existing donors and 2.1 million new donors between July 29 and August 27, 2014, as compared to just $2.7 million contributed to ALS for the same period in 2013 (ALS Association, 2014). Further research is needed to document whether this social media phenomenon was fueled by the flow experience of users or perhaps some other groupthink psychology.

Goal Clarity

For social media, clear goals can include sharing photos and video with friends and family, providing status updates, connecting with long-lost friends or family, broadening one's social network, and more. Consider the widely publicized May 7, 2014, instance of First Lady Michelle Obama tweeting a picture of herself holding a sign saying "#Bring Back our Girls" (Jackson, 2014). She sought to not only express her concern about the nearly 300 Nigerian girls who had been kidnapped by the terrorist group Boko Haram but also to help mobilize public support worldwide for the girls' safe return. This illustrates one type of goal that social media use might help the individual achieve.

Flexible and Appropriate Challenges

Challenges in social media include crafting messages that will interest others, taking photos or shooting videos that will capture attention, or reposting or curating interesting items. These challenges match well with the goals of communicating with one's social network and can flexibly accommodate varied situations or modalities, whether via a laptop computer, mobile device, or wearable gadget. Novice social media users may limit their activities to posting text messages with family and friends. Meanwhile, more experienced social media experts craft messages that combine text, photos, and videos to appeal to an audience of possibly millions of followers. Consider the case of the most popular Twitter feed as of May 23, 2014. Entertainer Katy Perry topped the Twittersphere on this date, with more than 53 million followers (TwitterCounter, 2014). A typical Perry post is relatively sophisticated, with a tweet from that date including text abbreviations, a link to a recent Instagram photo of her shoes, and a retweet from her Manchester show, not to mention a green heart graphic. Another example of a sophisticated Twitter user is U.S. Naval Commander Reid Wiseman, who has tweeted a number of photographs of the earth as seen from the International Space Station. These illustrate how social media can provide variable and appropriate levels of challenges to the individual user.

Immediate Feedback

Users post their content and in almost real time get feedback in the form of likes, comments, and reposting or retweeting of their posts. The art of conversation reveals much about flow. Conversation, whether via social media or face to face or a combination of both (e.g., Skype or Google Hangouts), is almost perfectly attuned to the potentialities of flow. Conversation is all about instant feedback, whether verbal or nonverbal. In a conversation via social media, two or more persons post, repost, and respond to each other's messages. As Mokros (1996) demonstrated, the interactive nature of conversation can play a fundamental role in identity, which may be a key factor in the transactional nature of social media and flow experience. Further research on this question may be revealing. It may provide a window into understanding how social media use may lead to changes in identity formation, change, and structure.

Alignment of Challenges and Skill

The requisite skills and challenges are generally well-aligned in the social media environment. When they are poorly aligned, the individual may feel frustration or boredom. When well aligned, the user experiences flow. Initially, the individual may learn to use a mobile device to take a picture or craft a message and how to share it via social networking media. Later she or he actively participates in the flow of social exchanges online. Over time, users can develop more advanced skills, or those with higher skill levels can seek greater challenges in the social networking space. For instance, an individual might start with text messaging. Eventually, she or he might start to post photos and then video, even editing videos for greater impact and reach. Texts formatted to Twitter require clarity of thought and attention to brevity, given the medium's 140-character limitation. Further, those who tweet are often motivated to increase their number of followers or reach even greater numbers by posting Tweets that are likely to be reposted. Witness the 2014 case of Academy Awards host Ellen DeGeneres and the widely retweeted selfie (photographic self-portrait) she took during the Oscar ceremony (Richford, 2014; Smith, 2014).

This alignment enables the individual to meet the challenges of a task. As a corollary, the requisite skills and challenges need to increase in parallel fashion over time. Otherwise, the user can experience either boredom when skills exceed the challenge or frustration when the challenge exceeds skills.

At the same time, when skills and challenges are not well aligned, negative psychological states can occur. These might include worry or apathy. Moreover, in the case of social media at least, a misalignment can even prove dangerous. Such was the case in early 2014 in India, when a 16-year-old teen from Kerala, India, attempted to take a selfie in front of a moving train, apparently in the hopes of posting the dramatic photo on a social media network. The boy stepped onto a set of train tracks holding a camera, hoping to get a selfie in front of the onrushing train, but he slipped and was run over by the train (Hazen, 2014). In Canada, another youth became a YouTube celebrity after posting video of himself getting kicked in the head and nearly killed (Gastaldo, 2014).

Autotelic Personality

Research also shows that flow states are enhanced or more likely to occur or be achieved if the individual has what is called an autotelic personality or develops her or his autotelic traits (Csikszentmihalyi, 1997). Autotelic means the individual derives enjoyment from the process or activity itself and not just the results that can be obtained from it. In other words, when the individual pursues an activity for the intrinsic reward or value of that activity, it facilitates achieving and maximizing the flow state. For instance, someone who enjoys taking photos for their own sake is more apt to achieve a flow state while sharing photos online than someone who takes the photos primarily for the likes he or she may get on Facebook. That is, process is more relevant than outcome, particularly if the outcome is delayed. However, it is important that the activity provide some sort of feedback along the way, although it may be quite varied in form. For instance, getting likes and seeing an increase in the number of one's followers will increase the intensity of the flow state.

In a broader media context, this means that people who enjoy the process of taking pictures, composing tweets, and sharing their messages are more likely to achieve a flow state than people who are primarily motivated to see how many times their messages might be retweeted or how many likes they can get on their Facebook posts or how many friends and followers they have, although these elements can contribute to flow. Yet this feedback can strengthen the flow state when it occurs.

Traditional media such as TV are passive and offer low-level order and structure to consciousness and attention but not much growth. Many TV viewers may spend hours a day watching television and are often left with a relatively empty feeling of having wasted their time, or they may feel bored or sometimes relaxed but unfulfilled. Television provides little opportunity for interactivity or participation and feedback, although this is beginning to change. As more viewers watch TV while utilizing a second screen, often a tablet or smartphone connected to the Internet, more viewers are utilizing social media or otherwise participating in media, sometimes related to the program being watched. Cole (2013) reported that worldwide, the typical person spends one third of awake time looking at a screen. More often than not it is a mobile device that supports active rather than passive media behavior: "90 per cent of teenagers sleep next to their mobile phone," Cole explained. "The obvious answer why is that it's their alarm clock." Yet, "another reason teens sleep beside their phones is because of FOMO, the fear of missing out." In other words, teens are increasingly engaged in 24-hour digital conversations, a process that occupies their mind even while sleeping. This means they are seeking to maintain a flow state.

Connecting to a Larger Whole

The seventh condition is perhaps the most intriguing. It suggests that an individual is especially likely to achieve a deep flow state if the activity helps her or him feel connected to a larger whole. For social media, this is a particularly likely condition, as sharing messages and other media with family, friends, or even larger communities can lead the individual to feel a direct link to a larger, even global community in virtually real time.

Concluding Reflections: A Flow Social Media Hypothesis

Social media have quickly taken a central place in the lives of more than a billion people around the world and show little signs of declining in popularity. Research offers a number of theoretical perspectives on why people are so drawn to the use of social media and the consequences of that engagement.

Among the most compelling theoretical perspectives is the psychological theory of flow, which refers to a positive mental state individuals can derive while engaged in an activity that meets certain conditions. These conditions include engaging in a purpose-driven activity (physical, mental, or both), where a confluence of challenges and skills are well aligned, feedback is frequent, and there is an opportunity to feel connected to a broader, larger purpose or community. Social media can meet these conditions. They can provide the individual with a purpose-driven and varied behavior where feedback is frequent, skills and challenges are well-aligned, and opportunities to make broader connections are widely available with the advent of networked mobile media.

As such, flow may explain much of the popularity of social media. The qualities of social networking media align remarkably well with the conditions that give rise to the positive psychological state of flow. In contrast to traditional media, which are largely passive for the consumer, social media engage the user actively, with purpose, matching challenges with skills and providing instant and often continuous feedback.

A key element that may even further explain the popularity of social media from a flow theory perspective hinges on the potential for individuals to use those networking tools to participate in or feel connected to something larger than themselves. In flow theory generally, becoming a part of a larger whole or at least obtaining a feeling of connection to a larger picture is a key element to achieving deep flow. Social media, as part of a global communication network, presents just such an opportunity for the individual to participate in a process of shared messages that are literally part of a larger, possibly worldwide network.

Further research can help to establish the extent to which social media users experience the flow state, when, and how often. Research can also help advance understanding of how and when emerging media, such as augmented and virtual reality, could facilitate the flow state, particularly deep flow. Research by Kim and Biocca (1997) and Steuer (1992) suggested that there are some positive psychological associations with telepresence, including memory and persuasion.

It is also vital to consider the ethical context for using social media and other emerging media to engage and nurture a flow state. Moreover, the health consequences, both positive and negative, associated with flow and the use of social media are important research topics. Avoiding addiction and advancing a positive health and social outcome are issues that should frame the understanding of flow and public social media engagement. Youth in particular may need guidance on the effective and safe parameters for social media use, especially as they intersect with the pursuit of flow states, whether intentional or not. Research shows that obsessive behavior is likely to lead to increased feelings of depression and other harmful health effects. Research is needed on how overinvolvement with social media may similarly lead to these same consequences. The potential value of integrating social media literacy and flow experience into school curricula is also a topic worth investigating. Flow offers the potential to enhance the quality of human experience, including via social media.

Consider the case of the nationwide, even global social media conversation that erupted after the unexpected death of actor and comedian Robin Williams in August 2014. Williams's sudden death prompted a spontaneous public conversation, especially via social media, exploring not only the actor and his career but also his suicide. Many millions of people shared thousands of messages about Williams through social media. As much as this event demonstrated the popularity of social media, it underscored the potential role that the flow state can play in a globally networked age.

As social media continue to evolve, particularly as new forms emerge to make them even more ubiquitous globally, research can help test the viability of a hypothesis linking flow to social media usage across time, sociocultural boundaries, and technological advance.

Note

The author thanks his wife, Jackie O. Pavlik, an artist, educator, and digital media authority, and Mari Assefa, a New York-based creative SEO (search engine optimization) and web developer, for lending their expertise in the realm of social media analytics.

References

Agence France Press. (2013, December 24). About 880 billion photographs will be taken in 2014—Including a lot of selfies. *Business Insider.* Retrieved January 15, 2015, from http://www.businessinsider.com/selfies-and-2013-2013-12

ALS Association. (2014, August 27). *Ice bucket donations continue to rise: $94.3 million since July 29.* Retrieved January 15, 2015, from http://www.alsa.org/news/media/press-releases/ice-bucket-challenge-082714.html

Ancu, M., & Cozma, R. (2009). Myspace politics: Uses and gratifications of befriending candidates. *Journal of Broadcasting & Electronic Media, 53*(4), 567–583.

Ballard, C. L. (2011). *"What's happening" @Twitter: A uses gratifications approach* (Master's thesis). University of Kansas, Lawrence. Retrieved January 30, 2015, from http://uknowledge.uky.edu/gradschool_theses/155

Carpentier, J., Mageau, G. M., & Vallerand, R. J. (2012). Ruminations and flow: Why do people with a more harmonious passion experience higher well-being? *Journal of Happiness Studies, 13*(3), 501–518.

CBS News. (2014, August 20). *How the "ice bucket challenge" is creating liquid gold.* Retrieved January 15, 2015, from http://www.cbsnews.com/videos/how-the-ice-bucket-challenge-is-creating-liquid-gold/

Chen, H., Wigand, R. T., & Nilan, M. S. (1999). Flow activities on the web. *Computers in Human Behavior, 15*(5), 585–608.

Chen, J. (2008). *Flow in games* (Master's thesis). University of Southern California, Los Angeles. Retrieved January 15, 2015, from http://www.jenovachen.com/flowingames/thesis.htm

Cole, J. (2013, March 12). *One-third of awake time looking at a screen.* Retrieved May 23, 2014 from http://aboutus.co.nz/dr-jeff-cole/

Csikszentmihalyi, M. (1990). *Flow: The psychology of optimal performance.* Oxford, UK: Oxford University Press.

Csikszentmihalyi, M. (1997). *Finding flow: The psychology of engagement with everyday life.* New York, NY: Basic Books.

Durkheim, E. (1893). *The division of labor in society.* New York, NY: Free Press.

Ellison, N. B., Steinfield, C., & Lampe, C. (2007). The benefits of Facebook "friends": Social capital and college students' use of online social network sites. *Journal of Computer-Mediated Communication, 12*(4), 1143–1168.

eMarketer. (2013, June 18). *Social networking reaches nearly one in four around the world.* Retrieved January 15, 2015, from http://www.emarketer.com/Article/Social-Networking-Reaches-Nearly-One-Four-Around-World/1009976

Gallion, A. J. (2010). *Applying the uses and gratifications theory to social networking sites: A review of related literature* (Unpublished paper). Purdue University, Fort Wayne, IN. Retrieved January 30, 2015, from http://www.academia.edu/1077670/Applying_the_ Uses_and_Gratifications_Theory_to_Social_Networking_Sites_A_Review_of_Related_ Literature

Gastaldo, E. (2014, April 22). *Viral video guy could get rich over kick in the head.* Retrieved January 15, 2015, from http://www.newser.com/story/185724/viral-video-guy-could-get-rich-over-kick-in-the-head.html

Ghani, J. A., & Deshpande, S. P. (1994). Task characteristics and the experience of optimal flow in human-computer interaction. *The Journal of Psychology, 128*(4), 381–391.

Ghani, J. A., Supnick, R., & Rooney, P. (1991). The experience of flow in computer-mediated and in face-to-face groups. In J. I. Degross, I. Benbasat, G. Desanctis, & C. M. Beath (Eds.), *Proceedings of the Twelfth International Conference on Information Systems, ICIS* (pp. 229–237), Atlanta, GA: Association for Information Systems.

Goleman, D. (1996). *Emotional intelligence: Why it can matter more than IQ*. London, UK: Bloomsbury Press.

Hazen, S. (2014, May 22). *Teen dies while trying to take selfie.* Retrieved January 30, 2015, from http://www.newser.com/story/187306/teen-dies-while-trying-to-take-selfie.html

Jackson, D. (2014, May 7). Michelle Obama tweets out support for Nigerian girls. *USA Today*. Retrieved January 15, 2015, from http://www.usatoday.com/story/theoval/ 2014/05/07/michelle-obama-nigeria-kidnapped-girls/8820621/

James, W. (1890). *The principles of psychology*. Cambridge, MA: Harvard University Press. Retrieved January 15, 2015, from http://psychclassics.asu.edu/James/Principles/ prin11.htm

Kim, T., & Biocca, F. (1997). Telepresence via television: Two dimensions of telepresence may have different connections to memory and persuasion. *Journal of Computer-Mediated Communication, 3*(2) [Online]. Retrieved April 25, 2015, from http://online library.wiley.com/doi/10.1111/j.1083-6101.1997.tb00073.x/full

Kotler, S. (2014). *The rise of Superman: Decoding the science of ultimate human performance*. Seattle, WA: Amazon Publishing/New Harvest.

Kraut, R., Lundmark, V., Patterson, M., Kiesler, S., Mukopadhyay, T., & Scherlis, W. (1998). Internet paradox: A social technology that reduces social involvement and psychological well-being? *American Psychologist, 53*(9) 1017–1031.

Kubey, R., & Csikszentmihalyi, M. (1990). *Television and the quality of life: How viewing shapes everyday experience*. Mahwah, NJ: Lawrence Erlbaum.

Kubey, R., & Csikszentmihalyi, M. (2002, February 2). Television addiction is no mere metaphor. *Scientific American*. Retrieved January 15, 2015, from http://www.commercial alert.org/issues/culture/television/television-addiction-is-no-mere-metaphor

Larson, R. (1988). Flow and writing. In M. Csikszentmihalyi & I. S. Csikszentmihalyi (Eds.), *Optimal experience: Psychological studies of flow in consciousness* (pp. 150–171). Cambridge, UK: Cambridge University Press.

Lunden, I. (2013, December 30). *73% of U.S. adults use social networks, Pinterest passes Twitter in popularity, Facebook stays on top*. Retrieved January 15, 2015, from http://techcrunch.com/2013/12/30/pew-social-networking/

Mageau, G. A., & Vallerand, R. J. (2007). The moderating effect of passion on the relation between activity engagement and positive affect. *Motivation and Emotion, 31,* 312–321.

Marr, A. J. (2001). *In the zone: A biobehavioral theory of the flow experience.* Retrieved January 15, 2015, from http://www.athleticinsight.com/Vol3Iss1/Commentary.htm

Mei, Y. (2012, July 2). *5 Chinese social networks you need to watch.* Retrieved January 15, 2015, from http://mashable.com/2012/07/02/china-social-networks/

Mirror. (2014, June 23). *The United States finally embraces the world's most popular game.* Retrieved January 15, 2015, from http://www.mirror.co.uk/sport/football/world-cup-2014/world-cup-2014-united-states-3746998

Mokros, H. B. (Ed.). (1996). *Interaction and identity: Information and behavior* (Vol. 5). Piscataway, NJ: Transaction Publishers.

Nakamura, J., & Csikszentmihalyi, M. (2001, December 20).Flow theory and research. In C. R. Snyder, E. Wright, & S. J. Lopez (Eds.), *Handbook of positive psychology* (pp. 195–206). Oxford, UK: Oxford University Press.

Nieoullon, A. (2002). Dopamine and the regulation of cognition and attention. *Progress in Neurobiology, 67*(1), 53–83.

Northwestern University in Qatar. (2014). *Entertainment use in the Middle East.* Retrieved January 15, 2015, from http://mideastmedia.org/online-and-social-media/introduction.html

Novak, T. P., & Hoffman, D. L. (1997). Measuring the flow experience among Web users. *Interval Research Corporation.* Retrieved April 25, 2015, from http://whueb.com/whuebiz/emarketing/research/m031121/m031121c.pdf

Novak, T. P., Hoffman, D. L., & Yung, Y-F. (1998). *Measuring the flow construct in online environments: A structural modeling approach* (Working paper, Owen Graduate School of Management, Vanderbilt University). Retrieved April 25, 2015, from http://www.academia.edu/2611185/Measuring_the_flow_construct_in_online_environments_a_struct ural_modeling_approach

Philippe, F. L., Vallerand, R. J., & Lavigne, G. (2009b). Passion makes a difference in people's lives: A look at well-being in passionate and non-passionate individuals. *Applied Psychology: Health and Well Being, 1*(1), 3–22.

Rathunde, K., & Csikszetnmihalyi, M. (2005). Middle school students' motivation and quality of experience: A comparison of Montessori and traditional school environments. *American Journal of Education, 111*(3), 341–371.

Richford, R. (April 8). *MIPTV: Ellen DeGeneres' Oscar selfie worth as much as $1 billion.* Retrieved January 15, 2015, from http://www.hollywoodreporter.com/news/miptv-ellen-degeneres-oscar-selfie-694562

Rousseau, F. L., & Vallerand, R. J. (2003). Le rôle de la passion dans le bien-être subjectif des aînés [The role of passion in the subjective well-being of elderly individuals]. *Revue québécoise de psychologie, 24*(3), 197–211. Retrieved January 15, 2015, from http://www.er.uqam.ca/nobel/r26710/LRCS/papers/121.pdf

Rousseau, F. L., & Vallerand, R. J. (2008). An examination of the relationship between passion and subjective well-being in older adults. *International Journal of Aging and Human Development, 66*(3), 195–211.

Schaffer, O. (2013). Crafting fun user experiences: A method to facilitate flow. *Human Factors International.* Retrieved January 15, 2015, from http://humanfactors.com/coolstuff/publications.asp

Seligman, M. E. P., & Csikszentmihalyi, M. (2000). Positive psychology: An introduction. *American Psychologist, 55,* 5–14.

Sherry, J. L. (2006). Flow and media enjoyment. *Communication Theory, 14*(4), 328–347.

Smith, C. L. (2014, March 2). Ellen DeGeneres' Oscars selfie beats Obama retweet record on Twitter. Retrieved January 15, 2015, from http://www.theguardian.com/film/2014/mar/03/ellen-degeneres-selfie-retweet-obama

Steuer, J. (1992). Defining virtual reality, *Journal of Communication, 42*(4), 73–93.

Stewart, J. (2000). Pathways to relapse: The neurobiology of drug- and stress-induced relapse to drug-taking. *Journal of Psychiatry Neuroscience, 25*(2), 125–136.

Tannenbaum, P. H. (1980). Entertainment as vicarious experience. In P. H. Tannenbaum (Ed.), *The entertainment functions of television* (pp. 107–131). New York, NY: Psychology Press.

TwitterCounter. (2014, May 23). Retrieved May 23, 2014, from http://twittercounter.com/katyperry

Urista, M. A., Dong, Q., & Day, K. D. (2008). Explaining why young adults use MySpace and Facebook through uses and gratifications theory. *Human Communication. A Publication of the Pacific and Asian Communication Association, 12*(2), 215–229.

Vallerand, R. J., Mageau, G. A., Elliot, A. J., Dumais, A., Demers, M. A., & Rousseau, F. L. (2008). Passion and performance attainment in sport. *Psychology of Sport and Exercise, 9*(3), 373–392.

Vallerand, R. J., Salvy, S. J., Mageau, G. A., Elliot, A. J., Denis, P., Grouzet, F. M. E., & Blanchard, C. B. (2007). On the role of passion in performance. *Journal of Personality, 75*(3), 505–534.

Webster, J., Trevino, L. K., & Ryan, L. (1993). The dimensionality and correlates of flow in human-computer interactions. *Computers in Human Behavior, 9*(4), 411–426.

Williams, R. (1974). *Television: Technology and cultural form.* London, UK: Fontana.

World Wide Web Foundation. (2008). *History of the Web.* Retrieved January 15, 2015, from http://webfoundation.org/visionandmission/history-of-the-web/

"For this much work, I need a Guild card!": Video Gameplay as a (Demanding) Coproduction

Nicholas David Bowman

"Games are a series of interesting decisions." (Meier, 2012)

Although there are likely as many perspectives as to what video games are as there are players, a central underlying theme of all video games is that they demand the player make decisions for the game to progress. To test this hypothesis, we need only to plug in and turn on our favorite video game console and advance to the start of actual gameplay (navigating start menus and options for continuing from previous game progression points and browsing past a litany of options to adjust game difficulty, color balance, and so forth), and then . . . simply do nothing. In most games, the gameworld will simply exist around the stationary avatar, executing its code and existing as if there were no other being, certainly not recognizing the player-avatar's presence. Other games—especially when combat oriented —will immediately recognize the player-avatar and set out to execute him/her/it as expeditiously as possible; without active resistance or avoidance by the player, such execution is likely swift.

On its surface, Meier's (2012) comment about games being "a series of interesting decisions" seems somewhat banal. After all, many flock to the medium largely because of the levels of autonomy and control—the number of interesting decisions available—not found in most entertainment media. Video games are what Jansz (2005) called a "lean-forward medium" which, through the near-constant stream of decisions required of the player, has the potential to bring players closer to both the digital/virtual (Tamborini & Skalski, 2006) and the narrative worlds (Green, Brock, & Kaufman, 2006) of the games, both of which have a substantial impact on game enjoyment.

However, a closer look at the series of interesting decisions gamers must make during their digital leisure might call into question the amount of work they are required to put into their entertainment experience. Interactive media are inherently unfinished texts; we have argued elsewhere that video games require active authorship by the player to be realized as complete texts (Bowman & Banks, in press). In this way, players are directly and indirectly recast as both on-screen agents and coproducers of their own digital experience. This burden of authorship is at once enjoyable and demanding, and in this chapter, I discuss the possible impact of the latter on the former.

Interactivity, Demand, and Coproduction

Although many definitions of interactivity can be found in the discourse surrounding video games, one of the more enduring is Steuer's (1992) definition of interactivity as the agency that a user has over the form and content of on-screen portrayals. Whether it be through the active manipulation of a game's controls (such as the analog game controller or the motion-capture active standard game controller to a more naturally mapped interface) or a game's content (such as the player-avatar's interactions with the on-screen characters and environments), each decision the player makes has direct and indirect consequences for what is shown on screen.

The roots of digital coproduction can be traced back to the roots of video games themselves, when a group of computer scientists and graduate students at MIT's Kluge Room (*kluge* is the German term for a clever contraption of ill-fitting parts with a particular purpose) included the user in the first guidelines for a computer demonstration program (an early term for what we now know as video games) in the 1960s (Graetz, 1981). The criteria included games designed to (a) demonstrate as many of a computer's resources as possible, (b) generate a new and unique run each time (to programmers, a run is each individual usage of a program), and (c) involve the onlooker "in a pleasurable and active way" (Graetz, 1981). At the most basic level, one can track the first board of *Super Mario Bros.* (1985, Nintendo) as an example of such coproduction. When the player initiates the start menu, she or he encounters a blue-sky world with a solid, flat terrain, entering stage-left with little guidance as to what lies ahead. Tapping the controller's directional pad to the right causes the player-avatar Mario to move unimpeded in that direction, until the sight of a similarly sized but angry entity (the Goomba) blocks the path, quickly closing in until a primal decision must be made: fight or flight? The decision tree is further complicated and expanded by the presence of shiny blocks marked with question marks that seem to beg to be explored further. Three contain coins and one contains a mushroom-shaped object that, when consumed, causes Mario to double in size and stature. This may encourage the player to reconsider an encounter with the Goomba or perhaps have more confidence to continue exploring the gameworld. Alternatively, players who ignore the tantalizing shiny boxes and trudge on ahead (regardless of whether they fight or flee the Goomba) are forced to take on a brave new world in which they are of equal stature to the other world denizens as they progress stage-right through the Mushroom Kingdom. Although both players—small and large—will experience the same kingdom, the former will be forced to progress with more trepidation than the latter, and the latter's size will afford greater access to other elements of the kingdom (e.g., the ability to shatter bricks for use as strategic platforms and to pick and eat fire flowers that allow the player to spit fire at on-screen enemies). All of these decisions are made within the first few minutes of play, and each has an impact on the game as realized by the player. Although the game's designers and producers have afforded the gamer the opportunity to make each decision, their agency is essentially removed at this stage, handed over to the player to decide how to author each step of the game—each press of the directional pad right paired with a press of the "B" or "A" button adds the player's own authorship to the *in situ* play experience. All of these decisions fit the Kluge Room's definition of a video game: *Super Mario Bros.* displayed a color palette and processing capability unmatched by any other home computing console at the time; given the decision tree discussed earlier, each individual run of a *Super Mario Bros.* level is completely dependent on the player's choices from the very first screen, and the game's popularity is one of the primary reasons for the Nintendo Entertainment System's market dominance throughout the 1980s and 1990s.

Today, *Super Mario Bros.* endures as nostalgic homage to the simpler times of the classic platform video game in which players progress left to right, collecting goods and navigating around obstacles along the way. Although much of this nostalgia is likely tied to the prominence that gaming experiences played for generations of today's working adults (Durkin, 2006), one might wonder whether the simpler times trope might also be connected to another development in video games: As games become more varied and intricate in the number of ways they can cognitively, affectively, behaviorally, and socially engage the player, they also become more demanding experiences in their own right. In other words, the rapid increase in coproduction can yield a demanding experience—insisting on our attention and cognition, tapping and triggering our emotions, guiding our behaviors and influencing how we interact with one another. Each of these is explored in detail here.

Attention and Cognitive Demand

Perhaps at the most basic level, video gameplay requires the player's attentional resources. On some level, such attentional demand is a prerequisite for the use of any form of communication—after all, if one doesn't attend to a message, it is difficult to process and respond to a message. However, one can argue that the attentional demand required to play a video game is greater than in other forms of entertainment such as television or film (Bowman & Tamborini, 2012, 2015). One reason for this increased attentional demand is the number of different cognitive abilities required to play a video game. For example, Bowman, Weber, Tamborini, and Sherry (2013) found that eye-hand coordination and mental rotational ability served as unique strong predictors of in-game performance at a first-person shooter—basing some of their findings on work correlating spatial ability skill with increased gameplay (Subrahmanyam & Greenfield, 1994). Green and Bavelier (2006) outlined several studies from their own research suggesting correlations between increased video gameplay and the ability to attend to and multitask among various visual stimuli.

Studies showing correlations between gaming and cognitive ability suggest, by proxy, that gaming is cognitively demanding. Evidence of this demand was provided by Nacke, Stellmach, and Lindley (2010), who used electroencephalogram data to demonstrate increased theta wave activity associated with the navigation of complex level designs such as games involving many landmarks and memory recalls. Increased brain activity was also associated with experiences of flow, suggesting that one way in which we can better understand players' loss of awareness when gaming (Sherry, 2006) might be associated with the manner in which some games absorb one's attentional resources. Work by Bowman and Tamborini (2012, 2015) further demonstrated not only that expectations of a game's attentional demand (conceptualized as task demand) can predict gamer's play preferences but also that attentional demand can predict mood; for example, increased attentional demand resulted in gamers' increased feelings of frustration.

Simply stated, the empirical record (including numerous studies not named earlier) has established that video games are an attentionally and cognitively demanding task—indeed, Grodal (2000) attributed the pleasure of gaming as an intended by-product of task, and Weber, Tamborini, Wescott-Baker, and Kantor (2009) defined media enjoyment as the synchronization of phasic attentional and reward networks. However, as games advance in their graphical, narrative, and technical complexity, they also advance in the variety of interesting decisions given to the player. As outlined in the limited capacity model for motivated mediated message processing (LM4CP: Lang, 2006), there is a general understanding that humans do not have an unlimited amount of cognitive ability, often devoting attentional resources to a stimulus only when given a reason to do so. The LM4CP is in line with the cognitive miser approach to social cognition (Fiske & Taylor, 1991), which suggests that humans conserve mental energy, treating it as a valuable resource to be expended only when necessary. Schwartz (2009) explored this further in his arguments about the inherently unsatisfying paradox of choice, suggesting that as the number of available choices goes up, one's satisfaction with any one of those choices decreases. Anyone who has ever saved several versions of a game just prior to a major decision in *Fable* or *Mass Effect* has been faced with the dilemma of "what might have been?" if the player would have properly defended the chicken coop from village thieves or chosen to brutally exact revenge for the murder of a comrade. We might even wonder whether this increased multitasking capacity might have a permanent impact on the brain. Loh and Kanai (2014) reported that individuals who engage in a high level of media multitasking (defined as "the concurrent consumption of multiple media forms") were statistically more likely to have less nerve density in the anterior cingulate cortex region of the brain, a region integral to cognitive and sensorimotor processing; in simpler terms, multitasking might be making our brains less functional.

For gamers, the task of gameplay is an incredibly cognitively demanding one, requiring the activation of any number of mental networks in order to coproduce and play through the experience. Gee (2003) argued that video games are by nature learning experiences, because players are so enmeshed in cocreating the interactive experience that they consistently learn and adapt as they play. In this way, video game production breaks from other forms of entertainment media in that it is not the role of the author, designer, or producer to complete the video game; rather, it is their collective role to set up a series of interesting decisions that allow the player to coproduce the experience by tapping into their own creative cognitive capacities (Bowman, Kowert, & Ferguson, in press), without frustrating or overtaxing those capacities.

Emotional Demand

Although the history of the medium is likely more rooted in cognitive rather than affective engagement—that is, pushing gamers to solve logic puzzles rather than to feel for the characters often trapped in those puzzles—video games have always been expected to engage the user's emotions on a deeper level than less-interactive lean-back forms of entertainment media. One reason for this expectation could be found in the ability of games to respond to players, representing an evolution in the medium from basic point-and-shoot to increasingly gripping emotional experience (Johnson, 2005). Game designer and scholar Chris Swain (as cited by Miller, 2013) explained that just as "film wasn't taken seriously as a medium until it learned to talk, games are waiting to learn to listen," and that this progression increasingly encourages players to consider carefully the consequences of their actions. MSNBC's Winda Benedetti wrote in 2010 about video games "peer(ing) into the dark reaches of the very real human heart to deliver stories that are thrilling, chilling and utterly absorbing" in describing what she and many others saw as a permanent trajectory for the video game as a storytelling vehicle poised to rival novels and films alike.

The emergence of the role-playing game (RPG) is likely most responsible for the video game's purposeful pull on the player's emotions. Hironobu Sakaguchi's *Final Fantasy*—released for the Nintendo Entertainment System in 1987—is a pivotal point in the evolution of video games. Written to represent the now-famous designer's final attempt to make a video game (hence the title), *Final Fantasy* was a narratively heavy video game that bucked the dominant trend by focusing on character development and conflict rather than on button-mashing and fast-paced action. Although many feared the game would flop (Fear, 2007), the title endures as one of the most popular on the Nintendo Entertainment System and sparked a franchise with over 170 individual game titles that have collectively sold over 1 billion copies worldwide (VGChartz, 2014).

Given the preponderance of violent video games readily available to players, one might question the relative focus on emotional attention required to pull a virtual trigger. Yet although familiarity with a video game seems to diminish guilt reaction in a shooting game (Hartmann & Vorderer, 2010), recent work using video games with more realistic scenarios—such as assuming the role of a terrorist in attacking innocent civilians in an airport takeover—elicited strong moral reactions related to guilt in players (Grizzard, Tamborini, Lewis, Wang, & Prabhu, 2014); similar guilt reactions were reported by Weaver and Lewis (2012) when playing *Fallout 3*. Such work is corroborated by Oliver et al. (2013), who found video games to be as capable of eliciting feelings of appreciation and meaningfulness in players as they are at eliciting feelings of fun and excitement—their work even found players recalling meaningful emotional experiences associated with playing more violent shooters such as *Call of Duty: Black Ops* and *Fallout 3* (a game that mixed RPG and first-person-shooter elements). Such results lend context to more recent theorizing in media psychology that video games are more than just stories with buttons (Elson, Breuer, Ivory, & Quandt, 2014).

Yet what might not be considered in this progression toward the emotionally serious game is a consideration for the player's role in the experience. How will audiences accustomed to gaming for escape and fantasy motivations (Sherry, Lucas, Greenberg, & Lachlan, 2006) cope with video games' demand on their emotional as well as attentional resources? In the aforementioned study by Grizzard et al. (2014), players of the game *Call of Duty: Modern Warfare 2* were faced with a mission in which they assumed the role of a terrorist operative, shooting children in an airport—an intentional programming choice designed to create an experience that "plays on [the player's] emotions . . . and, at times, maybe makes you feel some things you haven't felt in a game before" (Condrey, as cited by Takahashi, 2011). A similar experience can be found in *Spec Ops: The Line,* a similar type of shooter game (third person instead of first person) using the backdrop of a Middle Eastern war to challenge gamers' sense of morality, including a particularly poignant scene in which the player—after ordering a white phosphorous attack on an enemy column—later learns that the enemy troops were actually friendly forces transporting civilian refugees away from the war-torn city. Cocreator Walt Williams said that the purpose of the game was to force the players to judge themselves based on the results of their own on-screen actions (2013), that is, to challenge whether their actions can be considered heroic given their very real human toll. In an interview with Garland (2012), Williams explained that there are several endings to *Spec Ops: The Line* based on the decisions made by the player throughout the game—three of them canonical and involving various plot twists and turns revealing the short- and long-term psychological consequences of war on the protagonist, "And one in real life, for those players who decide they can't go on and put down the controller." Such a visceral reaction is in line with Bogost (2012), who discussed the potential for games—in particular, extremely violent games—to trigger specific disgust reactions in the player, placing an emotional toll that might serve to psychologically inhibit rather than encourage antisocial behavior.

The fact that gamers respond emotionally should not be a novel notion. As stated by Reeves and Nass (1996), the human mind has developed at a far slower rate than the technology available to it and, as a consequence, it is only natural for us to respond to on-screen emotional scenarios just as we would in the real world. Perhaps the difference now, as suggested by Oliver et al. (2013), is that whereas movie audiences might see sadness and emotion, in a video game, if a game character is crying, it is probably because the player both caused it and has some ability to rectify the situation. In this way, gamers coproduce the emotional tone of their games with each decision they make (Jansz, 2005) and find themselves in the unique position of being the catalyst of the hedonic valence of each gaming experience.

Behavioral Demand

Jansz's (2005) description of games as a lean-forward medium referred to the greater attentional and emotional draw of gaming compared to that of noninteractive media. Yet perhaps the most core component of video gameplay is the manner by which the players (users) input their desires into the game (interface) to cocreate the gaming experience. Playing a video game requires the player to take up both the physical and the metaphorical controller—assuming agency over some (if not all) aspects of the digital world and accepting the rights and responsibilities of that role.

As gamers take up their controllers, they immediately begin to form mental models, allowing them to cognitively (neo)associate their physical movements into on-screen actions. Skalski, Lange, and Tamborini (2006) demonstrated this by having players take up a motion sensor controller swung like a club when playing a golf video game or resembling a steering wheel when playing a driving game. When compared to the use of traditional handheld video game controllers, players using the motion sensor controls felt more spatially present in the digital environments—feeling more a part of the virtual world (Tamborini & Skalski, 2006)—which led to increased feelings of enjoyment. One explanation for the increased enjoyment using the motion sensor controls might have been that those controllers allowed players to use their existing real-world mental models rather than having to learn unique models for the video games, such as pressing and holding a trigger button to swing a golf club or accelerate an automobile; these results were confirmed in replication by McGloin, Farrar, and Krcmar (2011). Biocca (1997) referred to the similarity between real and on-screen actions as natural mapping, suggesting that interfaces that are more naturally mapped to analogous real-world behaviors are easier for players to use.

However, although these studies might lead us to suggest that the creation and refinement of naturally mapped control systems would yield a less-demanding (and therefore more enjoyable and presence-inducing) gaming experience, other studies have found natural mapping to be a perceptual rather than a technological variable. For example, Rogers, Bowman, and Oliver (2015) found that when playing a first-person shooter using either a motion sensor controller (a Wiimote) or a traditional controller (a Nintendo GameCube pad), players perceived the GameCube as more naturally mapped (in part because it was more aligned with cognitive skill sets associated with video gameplay) . The study questions the assumption that players can take up motion controllers without challenge and immediately map their mental models to those controllers. It may be the case that experienced gamers have established mental models associated with game controllers—similar to the experienced typist's reliance on the QWERTY keyboard layout, despite studies showing the century-old format to be suboptimal for computer keyboard operation (Noyes, 1983). For many experienced users, there is nothing more demanding than having to learn a new mental model in order to synchronize our eye-hand and cognitive systems toward an in-game action. Moreover, these motion-capture controllers assume by design that the gamer has the physical ability to engage the controller: For example, a camera-based dancing video game assumes the gamer has the physical ability and capacity to dance, which might unintentionally exclude some populations, such as gamers with disabilities.

Some scholars have connected in-game behaviors to out-of-game habits, suggesting a correlation between physical and ephemeral actions. Lange, Banks, and Lange (2014) found that an individual's real-world physical activity habits—such as proclivity toward walking when commuting to work—were significant predictors of players choosing to have their characters walk in a video game (compared to other modes of transportation, such as riding bicycles or driving a car). This association only held when the game provided players with a concrete objective, such as making a delivery to a discrete location. Research associating habitual behaviors and gaming is nascent, but the line of inquiry suggests that gamers might engage their own implicit social and behavioral norms when playing video games in an effort to reduce the cognitive load placed on them while gaming. With this in mind, game designers hoping to engage the player as an active coproducer of content might need to consider ways to disrupt such automated responses, shifting players away from and out of their habitual responses and guiding them to higher level decision-making behaviors, encouraging them to switch cognitive gears (Louis & Sutton, 1991) so that players will continue to lean forward (not backward) into the virtual space.

Players' reliance on habits is not necessarily restricted to their behavioral norms. For example, work on the impact of morality and decision making (Joeckel, Bowman, & Dogruel, 2012; Weaver & Lewis, 2012) has demonstrated that an individual's innate sense of morality plays an important biasing role in whether he or she will instruct (or control) others to behave in a moral or immoral manner. Specifically, Joeckel et al. (2012) reported what they saw as a moral versus amoral dichotomy—that is, in-game decisions were seen as either moral *gut* reactions or strategic and playful *game* decisions. For example, players with a sensitivity toward principles of care and harm were significantly less likely to harm others (or allow others to be harmed) in a role-playing game. Conversely, players whose moral sensitivities toward a particular domain were low did not behave in an immoral manner (i.e., cause harm to be inflicted), but rather, their observed in-game decisions were a matter of statistical chance—split 50/50 between violating and upholding the moral domain in question. Although the latter finding was not meant to suggest that players' observed video game decisions are simply the result of some random pressing of buttons, the former was interpreted as evidence that games hoping to engage a player's deliberate moral reasoning (such as the war shooters mentioned in the previous section on emotional demand) might simply trigger innate moral reactions. To apply the example of *Spec Ops: The Line*, data from these studies might suggest that whereas Williams's goal was to create a game that, through the portrayal of contextualized violence, allows players to judge themselves for the consequences of their actions (rather than allowing the game to judge them), games offering players too much latitude of behavioral choice might simply result in a coproduced experience that, through the innately biased decisions of the player, simply aligns with rather than challenges the player's moral code. Explorations of this dual-process moral system—by which one makes an instantaneous intuitive decision before reappraising that decision—have been guided by Tamborini's (2011) moral intuition and media entertainment model (such reappraisals underpinned the guilt reactions reported by Grizzard et al., 2014), but future work is needed to better understand the extent to which gamers exercise automatic and deliberate behaviors when playing inside virtual worlds.

Socially Demanding

Harkening back to the earliest days of *Spacewar!*—which had two input devices so that two players could do battle—video games have largely been designed as social technologies. Stretching back to the Atari VCS home game console (1977), one is hard pressed to find a video game console that did not include inputs for at least two game controllers and a barrage of games designed to support colocated gaming (Bowman, Weber, Tamborini, & Sherry, 2013). Studies of the uses and gratifications of video gameplay suggest that, just after challenge and competition motivations, the need for social interaction is a core gratification sought (Colwell & Kato, 2005; Sherry et al., 2006; Wan & Chiou, 2006).

A form of gaming particularly well suited for social interaction is the massively multiplayer online game, in which players have the opportunity to interact with countless other player-avatars in vast digital worlds (Cole & Griffiths, 2007; Yee, Ducheneaut, & Nelson, 2012). Huh and Bowman (2008) found that players with high levels of self-reported extroversion were drawn to playing games such as *World of Warcraft (WoW)* and argued that such games were well suited to meet the (hyper?)social demands of outgoing individuals. Such work corroborates Kowert and Oldmeadow's (2013, 2014) findings that not only do gamers forge and strengthen social bonds through their online play, but these online spaces can also serve to accommodate gamers who might not feel comfortable socializing in their daily lives. For these players, the facts that these interactions can happen both through a chosen avatar and in a chosen environment seem to mitigate their otherwise apprehensive tendencies. Put another way, shy gamers might find digital interactions to be less socially demanding and, as a result, open themselves up to communication with others—a specific finding reported in Kowert, Domahidi, and Quandt (2014).

Cooperative gaming can be a taxing experience but a rewarding one. One such reward of social gameplay is the development of transactive memory (TMS): a group's collective mechanism for reading, storing, and retrieving unique knowledge of a given situation (Wegner, Giuliano, & Hertel, 1985). Kahn (2012) conducted a series of studies investigating TMS in video games, using server data from some 16,000-plus *League of Legends* players to demonstrate how the mechanics of group selection and team size as well as communication channel modality and social information processing led to the formation of TMS. Follow-up work by Kahn and Bowman (2014) found that increased perceptions of task demand (a proxy measure for the perceived difficulty of in-game tasks, such as discrete battle instances) led to increases in TMS and that this relationship strengthened over time. Related to transactional memory and team building is the players' ability to produce their own social capital as a function of their gaming success, acquiring accolades in the form of gaming gear, in-game knowledge, game skill, and membership in prestigious gamer groups and guilds (Consalvo, 2009).

The sociality of video games is receiving attention from researchers (Quandt & Kroger, 2014), but the push for increased social interaction between gamers might bring with it an increase in social demand—a variable that has not received much empirical consideration. For example, a prominent feature of online gaming is how they are played in groups, which often brings with it ad hoc social roles. Chen (2011) observed a group of elite *WoW* players and found an incredible amount of labor put into the experience by more senior players offering tutelage to younger ones. In-depth interviews of nearly three dozen *WoW* players conducted by Banks (2013) suggested that as the game became more socially demanding, players seemed to withdraw somewhat from the sociality of the event and instead approached it as a task event—using the player-avatars as a means to meet some in-game end goal. Following up on this work and focusing specifically on the player-avatar relationships (PARs), Banks (2013) reported a range of PARs from avatar-as-object to avatar-as-me and avatar-as-other. The avatar-as-other relationship was an important finding in that it revealed that players at times place the locus of control in the avatar's hands rather than their own. Players in that study recognized the avatar's moral values as well as its/his/her intrinsic motivations to fight, love, or wonder about the deliberations of its/his/her world. Linguistic analysis of these interviews (Banks & Bowman, in press) found avatar-as-other players were significantly more likely to refer to their avatar in the third person, discussing it as a distinct social other with its own influence on the PAR. From a PAR perspective, one might challenge the extent to which the player has full control over the coproduction of an interactive narrative when the avatar has a distinct and legitimate role. Put simply, avatars have their own needs, and players must respect those needs in their gameplay practice.

The Input Is the Experience

Because video games will always require input from the user, we might suggest that we consider the human element of the video game—not the creation of human-like characters but rather, the "squishy bit" on the other side of the screen: the human player himself or herself. From the first

Spacewar! (1962) to the most recent *Super Mario 3D World* (2013), video games have presented gamers with a series of interesting decisions to make for more than 50 years. The development of the technology has made it possible for players to have equally enjoyable and meaningful digital experiences by involving—even taxing—players at unprecedented cognitive, affective, behavioral, and social levels. The fruits of this coproduction-*qua*-gameplay are experiences that challenge the brain, heart, hands, and bonds with others in ways that have the potential to, as suggested by Christakis and Fowler (2009), bring us closer to our own humanity—even in an age of persistent mediation.

Gameography

Call of Duty: Black Ops. (2010). Designed by Corky Lehmukuhl, David Vonderhaar, & Joe Chiang. Written by Craig Houston, Dave Anthony, & David S. Goyer.

Call of Duty: Modern Warfare 2. (2009). Designed by Todd Alderman, Steve Fukuda, Mackey McCandlish, Zied Rieke, Mohammad Alvai, & Jon Porter. Written by Jesse Stern.

Fable. (2004). Designed by Peter Molyneux.

Fallout 3. (2008). Designed by Emil Pagliarulo & Joel Burgess. Written by Emil Pagliarulo.

Final Fantasy. (1987). Designed by Hironobu Sakaguchi, Hiromichi Tanaka, Akitoshi Kawazu, & Koichi Ishii. Written by Horonubu Sakaguchi, Kenji Terada

League of Legends. (2009). Designed by Christina Norman, Rob Garrett, & Steve Feak. Produced by Steven Snow & Travis George.

Mass Effect. (2007). Published by Microsoft Game Studies.

Spacewar! (1962). Programmed by Steve Russell.

Spec Ops: The Line (2012). Designed by Cory Davis. Written by Walt Williams & Richard Pearsey.

Super Mario Bros. (1985). Programmed by Toshihiko Nakago & Kazuaki Morita. Designed by Shigeru Miyamoto & Takashi Tezuka.

Super Mario 3D World. (2013). Produced by Yoshiaki Koizumi. Published by Nintendo.

World of Warcraft. (2004). Designed by Rob Pardo, Jeff Kaplan, & Tom Chilton. Published by Blizzard Entertainment.

References

Banks, J., & Bowman, N. D. (2014). The language of our player-avatar relationships: Self-differentiation as key to understanding para-social and social interactions with video game characters. New Media & Society. retrieved from http://nms.sagepub.com/content/early/2014/10/16/1461444814554898.full.pdf+html

Banks, J. D. (2013). *Human-technology relationality and self-network organization: Player and avatars in World of Warcraft* (Unpublished doctoral dissertation). Colorado State University, Fort Collins.

Benedetti, W. (2010, May 6). Video games get real and grow up. *MSNBC*. Retrieved January 6, 2015 from http://www.msnbc.msn.com/id/36968970/ns/technology_and_science-games/#.UJE8PqNCesY

Biocca, F. (1997). The cyborg's dilemma: Progressive embodiment in virtual environments. *Journal of Computer-Mediated Communication, 3*(2). [Online] Retrieved April 25, 2015 from http://onlinelibrary.wiley.com/doi/10.1111/j.1083-6101.1997.tb00070.x/full

Bogost, I. (2012). Disinterest. In I. Bogost (Ed.), *How to do things with video games* (p. 134–140). Minneapolis: University of Minnesota Press.

Bowman, N. D., & Banks, J. (in press). Playing the zombie author: Machinima through the lens of Barthes. In K. Kenney (Ed.), *Philosophy for multisensory communication*. New York, NY: Peter Lang.

Bowman, N. D., Kowert, R., & Ferguson, C. (in press). Powering up the creative mind: The impact of video game play and violent content. In G. Green & J. Kaufman (Eds.), *Powering up the creative mind*. New York, NY: Academic Press.

Bowman, N. D., & Tamborini, R. (2012). Task demand and mood repair: The intervention potential of computer games. *New Media & Society, 14*(8), 1339–1357.

Bowman, N. D., & Tamborini, R. (2015). "In the mood to game": Selective exposure and mood management processes in computer game play. *New Media & Society, 17*(3), 375–393.

Bowman, N. D., Weber, R., Tamborini, R., & Sherry, J. L. (2013). Facilitating game play: How others affect performance at and enjoyment of video games. *Media Psychology, 16*(1), 39–64.

Chen, M. (2011). Leet Noobs: *The life and death of an expert player group in World of Warcraft*. New York, NY: Peter Lang.

Christakis, N. A., & Fowler, J. A. (2009). *Connected: The surprising power of social networks and how they shape our lives*. New York, NY: Little, Brown.

Cole, H., & Griffiths, M. (2007). Social interactions in massively multiplayer online role-playing games. *CyberPsychology & Behavior, 10*(4), 575–583.

Colwell, J., & Kato, M. (2005). Video game play in British and Japanese adolescents. *Simulation & Gaming, 36,* 518–530.

Consalvo, M. (2009). *Cheating: Gaining advantage in videogames*. Cambridge, MA: The MIT Press.

Durkin, K. (2006). Game playing and adolescents' development. In P. Vorderer & J. Bryant (Eds.), *Playing computer games: Motives, responses, and consequences* (pp. 415–428). Mahwah, NJ: Lawrence Erlbaum.

Elson, M., Breuer, J., Ivory, J. D., & Quandt, T. (2014). More than stories with buttons: Narrative, mechanics, and context as determinants of player experience in digital games. *Journal of Communication, 64*(3), 521–542.

Fear, E. (2007, December 13). *Sakaguchi discusses the development of Final Fantasy*. Develop. Intent Media. Retrieved April 25, 2015 from http://www.develop-online.net/news/sakaguchi-discusses-the-development-of-final-fantasy/0102088

Fiske, S. T., & Taylor, S. E. (1991). *Social cognition: From brains to culture* (2nd ed.). New York, NY: McGraw-Hill.

Garland, J. (2012, July 16). Aftermath: Crossing the line with Walt Williams. *GamingBolt.com*. Retrieved, January 6, 2015 from http://gamingbolt.com/aftermath-crossing-the-line-with-walt-williams

Gee, J. (2003). *What video games have to teach us about learning and literacy*. New York, NY: Palgrave Macmillan.

Graetz, J. M. (1981, August). The origin of Spacewar! *Creative Computing Magazine*. Retrieved February 27, 2015 from http://www.wheels.org/spacewar/creative/SpacewarOrigin.html

Green, C. S., & Bavelier, D. (2006). The cognitive neuroscience of video games. In P. Messaris & L. Humphreys (Eds.), *Digital media: Transformations in human communication* (pp. 211–224). New York, NY: Peter Lang.

Green, M. C., Brock, T. C., & Kaufman, G. F. (2006). Understanding media enjoyment: The role of transportation into narrative worlds. *Communication Theory, 14*(4), 311–327.

Grizzard, M., Tamborini, R., Lewis, R. J., Wang, L., & Prabhu, S. (2014). Being bad in a video game can make us morally sensitive. *CyberPsychology, Behavior, and Social Networking, 17*(8), 499–504.

Grodal, T. (2000). Video games and the pleasures of control. In D. Zillmann & P. Vorderer (Eds.), *Media entertainment: The psychology of its appeal* (pp. 197–213). Mahwah, NJ: Lawrence Erlbaum.

Hartmann, T., & Vorderer, P. (2010). It's okay to shoot a character: Moral disengagement in violent video games. *Journal of Communication, 60*(1), 94–119.

Huh, S., & Bowman, N. D. (2008). Perception and addiction of online games as a function of personality traits. *Journal of Media Psychology, 13*(2), 1–31 [online]. Retrieved January 6, 2015 from http://www.calstatela.edu/faculty/sfischo/Bowman%20online%20game%20addiction-final.doc

Jansz, J. (2005). The emotional appeal of violent video games for adolescent males. *Journal of Communication, 15*(3), 219–241.

Joeckel, S., Bowman, N. D., & Dogruel, L. (2012). Gut or game: The influence of moral intuitions on decisions in virtual environments. *Media Psychology, 15*(4), 460–485.

Johnson, S. (2005). *Everything bad is good for you*. New York, NY: Riverhead.

Kahn, A. (2012). *We're all in this (game) together: Transactive memory systems, social presence, and social information processing in video game teams* (Unpublished doctoral dissertation). Pasadena: University of Southern California.

Kahn, A., & Bowman, N. D. (2014, May). *With tough work comes tough responsibility: The association between perceived task demand and transactive memory in video game teams*. Paper presented at the annual meeting of the International Communication Association, Seattle, WA.

Kowert, R., Domahidi, E., & Quandt, T. (2014). The relationship between online video game involvement and game-related friendships among emotionally sensitive individuals. *CyberPsychology, Behavior, and Social Networking, 17*(7), 447–453.

Kowert, R., & Oldmeadow, J. A., (2013). (A)Social reputation: Exploring the relationship between online video game involvement and social competence. *Computers in Human Behavior, 29*(4), 1872–1878.

Kowert, R., & Oldmeadow, J. A. (2014). Playing for social comfort: Online video game play as a social accommodator for the insecurely attached. *Computers in Human Behavior* [online]. Retrieved April 25, 2015 from http://www.researchgate.net/publication/262068867_Playing_for_social_comfort_Online_Video_Game_Play_as_a_Social_Acco mmodator_for_the_Insecurely_Attached

Lang, A. (2006). The limited capacity model of mediated message processing. *Journal of Communication, 50*(1), 46–70.

Lange, R., Banks, J., & Lange, A. (2014, May). *The influence of physical activity habits on observed video game travel mode decisions.* Paper presented at the annual convention of the International Communication Association, Seattle, WA.

Loh, K. K., & Kanai, R. (2014). Higher media multi-tasking activity is associated with smaller gray-matter density in the anterior cingulate cortex. *PLoS One.* Retrieved February 27, 2015 from http://journals.plos.org/plosone/article?id=10.1371/journal.pone.0106698

Louis, M. R., & Sutton, R. I. (1991). Switching cognitive gears: From habits of mind to active thinking. *Human Relations, 44*(1), 55–76.

McGloin, R., Farrar, K. M., & Krcmar, M. (2011). The impact of controller naturalness on spatial presence, gamer enjoyment, and perceived realism in a tennis simulation video game. *Presence: Teleoperators and Virtual Environments, 20*(4), 309–324.

Meier, S. (2012, March). *Interesting decisions.* Presentation at the Game Developers Conference, San Francisco, CA.

Miller, P. (2013, March 27). Jessee Schell's search for the Shakespeare of video games. *Gamasutra.* Retrieved January 6, 2015 from http://www.gamasutra.com/view/news/189370/Jesse_Schells_search_for_the_Shakespeare_of_video_games.php

Nacke, L. E., Stellmach, S., & Lindley, C. A. (2010). Electroencephalographic assessment of player experience: A pilot study in affective ludology. *Simulation & Gaming, 42*(5), 632–655.

Noyes, J. (1983). The QWERTY keyboard: A review. *International Journal of Man-Machine Studies, 18*(3), 265–281.

Oliver, M. B., Bowman, N. D., Woolley, J., Rogers, R., Sherrick, B., & Chung, M-Y. (2013, June). *Video games as meaningful entertainment experiences.* Paper presented at the annual meeting of the International Communication Association, London, UK.

Quandt, T., & Kroger, S., (2014). *Multiplayer: The social aspects of digital gaming.* New York, NY: Routledge.

Reeves, B., & Nass, C. (1996). *The media equation: How people treat computers, television, and new media like real people and places.* Cambridge, UK: Cambridge University Press.

Rogers, R., Bowman, N. D., & Oliver, M. B. (2015). It's not the model that doesn't fit, it's the controller! The role of cognitive skills in understanding the links between natural mapping, performance, and enjoyment of console video games. *Computers in Human Behavior, 49*, 588-596.

Schwartz, B. (2009). *The paradox of choice.* New York, NY: HarperCollins.

Sherry, J. L. (2006). Flow and media enjoyment. *Communication Theory, 14*(4), 328–347.

Sherry, J. L., Lucas, K., Greenberg, B., & Lachlan, K. (2006). Video game uses and gratifications as predicators of use and game preference. In J. Bryant & P. Vorderer (Eds.), *Playing video games: Motives, responses, and consequences* (pp. 213–224). Mahwah, NJ: Lawrence Erlbaum.

Skalski, P., Lange, R., & Tamborini, R. (2006). Mapping the way to fun: The effect of video game interfaces on presence and enjoyment. *Proceedings of the Ninth Annual International Workshop on Presence.* Cleveland, OH: Cleveland State University.

Steuer, J. (1992). Defining virtual reality: Dimensions determining telepresence. *Journal of Communication, 4*(4), 73–93.

Subrahmanyam, K., & Greenfield, P. M. (1994). Effect of video game practice on spatial skills in girls and boys. *Journal of Applied Developmental Psychology, 15*(1), 13–32.

Takahashi, D. (2011, November 7). Modern Warfare 3's disturbing scene involves child's death. *VentureBeat.com.* Retrieved January 6, 2015 from http://venturebeat.com/2011/11/07/modern-warfare-3s-disturbing-scene-involves-childs-death/

Tamborini, R. (2011). Moral intuition and media entertainment. *Journal of Media Psychology, 23*(1), 39–45.

Tamborini, R., & Bowman, N. D. (2010). Presence in video games. In C. Bracken & P. Skalski (Eds.), *Immersed in media: Telepresence in everyday life* (pp. 87–110). New York, NY: Routledge.

Tamborini, R., & Skalski, P. (2006). The role of presence in the experience of electronic games. In P. Vorderer & J. Bryant, (Eds.), *Playing video games: Motives, responses, and consequences* (pp. 225–240). Mahwah, NJ: Lawrence Erlbaum.

VGChartz.com. (2014, September 27). Game database: Final Fantasy. Retrieved January 6, 2015 from http://www.vgchartz.com/gamedb/?name=Final+Fantasy

Wan, C. S., & Chiou, W. B. (2006). Psychological motives and online game addiction: A test of flow theory and humanistic need theory for Taiwanese adolescents. *CyberPsychology & Behavior, 9,* 317–324.

Weaver, A. J., & Lewis, N. (2012). Mirrored morality: An exploration of moral choice in video games. *CyberPsychology, Behavior, and Social Networking, 15*(11), 610–614.

Weber, R., Tamborini, R., Wescott-Baker, A., & Kantor, B. (2009). Theorizing flow and media enjoyment as cognitive synchronization of attentional and reward networks. *Communication Theory, 19*(4), 397–422.

Wegner, D. M., Giuliano, T., & Hertel, P. (1985). Cognitive interdependence in close relationships. In W. J. Ickes (Ed.), *Compatible and incompatible relationships* (pp. 253–276). New York, NY: Springer-Verlag.

Williams, W. (2013, March). *We are not heroes: Contextualizing violence through narrative.* Presentation at the Game Developers Conference, San Francisco, CA. Retrieved January 6, 2015 from http://www.gdcvault.com/play/1017980/We-Are-Not-Heroes-Contextualizing

Yee, N., Ducheneaut, N., & Nelson, L. (2012). Online gaming motivations: Development and validation. *Proceedings of the SIGCHI Conference on Human Factors in Computing Systems,* 2803–2806.

The Mobile Conversion, Internet Regression, and the Repassification of the Media Audience

Philip M. Napoli and Jonathan A. Obar

In late 2010, mobile broadband Internet subscriptions overtook the number of subscriptions over fixed technologies worldwide (Bold & Davidson, 2012). This growth is expected to continue, with projections suggesting that mobile subscriptions will rise from 61% of all broadband connections in developing countries in 2012 to 84% in 2016 (Bold & Davidson, 2012).

There are a number of important dimensions of this ongoing diffusion of mobile Internet access. First, it is important to recognize that, for an increasing proportion of the population worldwide, mobile-based forms of Internet access represent the *primary* means of going online (see Napoli & Obar, 2014). Furthermore, for some sectors, mobile-based forms of Internet access are the *only* means for connecting online. According to one recent estimate, there were approximately 14 million mobile-only Internet users in the world in 2011, with the number expected to increase to 788 million by 2016 (Horner, 2011). Naturally, these patterns are accompanied by declines associated with more traditional forms of Internet access (Napoli & Obar, 2014).

Much has been written about the tremendous benefits, and even the transformative capacity, associated with this global mobile diffusion (e.g., Castells, Fernandez-Ardevol, Qiu, & Sey, 2007). The rapid, global diffusion of smartphones provides users with unprecedented levels of mobile information access and communicative capacity. The benefits of this diffusion of mobile forms of Internet access are manifest not only in terms of expanding the contexts in which Internet use is possible but also in terms of providing those who previously could not afford Internet access (due to the greater costs associated with traditional, fixed forms of Internet access) with a more affordable means of getting online. There is compelling evidence across a variety of contexts that mobile Internet access can provide those without traditional forms of Internet access with opportunities to become better integrated into social, economic, and political life (Chigona, Beukes, Vally, & Tanner, 2009; Chircu & Mahajan, 2009; Fong, 2009; Schejter & Tirosh, 2012; Wareham, Levy, & Shi, 2004). A convincing case can be made that, in many ways, mobile Internet access is superior to traditional Internet access, particularly in terms of the wider array of contexts in which it enables access and different kinds of uses (Arthur, 2011; Horrigan, 2009; Purcell, Rainie, Rosenstiel, & Mitchell, 2011).

Although there is much to be gained from what might be called the on-going *mobile conversion,* in which mobile Internet access supplants wireline access via PCs/laptops, there are significant drawbacks as well—which have received significantly less analytical attention than the benefits. Consequently, we offer a somewhat contrarian perspective to the overwhelmingly positive discourse that has accompanied discussions of the rise of mobile Internet access. Specifically, we argue that the transition from fixed to mobile forms of Internet access represents an evolutionary regression across some key dimensions. In particular, the mobile conversion brings with it a significant step backward in terms of the activity and autonomy that the Internet has, to this point, brought to media audiences. It is this repassification[1] of the audience that is the focal point of this analysis.

In addressing these issues, we begin by examining the institutional tensions and resistance patterns that have historically characterized the dynamic between media and audiences, with a particular emphasis on the extent to which media systems have facilitated or discouraged audience activity and content creation. Next, we examine the evidence that the dynamics of mobile Internet content provision and usage are fundamentally different from the traditional PC-based Internet[2] in ways that represent a regression of the Internet's capabilities, particularly in terms of facilitating a more active, content-creating, and distributing role for the audience. Relative to the PC-based model of Internet access and usage, the mobile Internet is, in many ways, a significant step back toward a more passive audience model in which the traditional boundaries between content providers and audiences that the Internet has thus far helped to break down are to some extent being reestablished.

Media/Audience Evolution

A key dimension of the history of media has been the extent to which different technological forms, and the different institutionalized conventions surrounding these forms, either facilitate or discourage audience activity and participation (Napoli, 2011). Certainly one of the defining aspects of the emergence of the Internet, the web, and later, social media applications, has been the extent to which they facilitate greater audience activity, engagement, and participation in the production and distribution of media content than was the case with legacy mass media such as print, broadcasting, and cable (Cover, 2004; Harrison & Barthel, 2009). The rise of the networked information economy and online forms of commons-based peer production (Benkler, 2006), the growth of a digitally mediated fifth estate (Benkler, Roberts, Faris, Solow-Niederman, & Etling, 2013; Dutton, 2009; Obar & Shade, 2014), the increasing popularity of digital forms of political activism (Obar, 2014; Obar, Zube, & Lampe, 2012), and the growth of virtual worlds for socializing, gaming, and other forms of online interaction (Castronova, 2001; Lastowka, 2010) are just a few examples from the burgeoning multitude of activities suggesting that audiences are increasingly participating and engaging with media technologies and content online.

It has been well documented that the diffusion of the Internet has facilitated a transformation of the audience from "passive observer to active participant in a virtual world" (Livingstone, 2003, p. 338; see also Svoen, 2007). The position of audiences on the active-passive continuum naturally connects with the other important vector of audience evolution facilitated by the Internet—the distinction between media producers and distributors and media consumers. A substantial amount of scholarly attention over the past decade has been devoted to the ways in which the boundaries separating content producers/distributors and content consumers have become increasingly blurred by digital media technologies (see, e.g., Lunenfeld, 2011). The key point in terms of this analysis is the extent to which the platforms via which individuals access and use the Internet have facilitated these transformations.

What is particularly interesting about this aspect of the evolution of the media audience in the Internet age is that it represents, in some ways, a return to a formulation of the audience that was predominant in the pre-mass media era. The literature on the history of audiences has frequently demonstrated that early manifestations of the audience were very much participatory and interactive (Billings, 1986; Butsch, 2008; Harrison & Barthel,

2009). Theater audiences, for example, once engaged in a wide array of activities, ranging from singing songs to yelling instructions and insults at the performers to yelling at (and fighting with) each other (Billings, 1986; Butsch, 2008). Even in the developmental days of some of the mass media, elements of an active, participatory audience persisted (Griffen-Foley, 2004; Newman, 2004). Ross (1999), for instance, chronicled the worker film movement in the early years of the 20th century, in which members of various workers' organizations, frustrated with the antilabor messages they perceived in motion pictures, set about producing and distributing their own films. In the early days of the radio industry, audiences (primarily via fan mail) played an integral role in the creation of radio programs, as producers and writers frequently incorporated suggestions received from audience members into their scripts (Razlogova, 1995).

However, as the electronic mass media developed and evolved, the dynamic between content provider and audience became increasingly unidirectional, and the more familiar one-to-many, top-down content provider-audience dynamic within these media became institutionalized (Butsch, 2008; Razlogova, 1995). From this standpoint, one could characterize the traditional mass media as contributing to a *passification* of audiences, with this tendency toward passivity becoming, to some extent, a defining element of the very idea of the audience (Butsch, 2008).

As noted earlier, over the past two decades, due primarily to the diffusion of the Internet, the more unidirectional approach to the relationship between media and audiences that took hold during the growth of traditional mass media through the 20th century has begun to be undone. The Internet has helped to "effectively *restore* to the audience their capacity to participate" (Cover, 2006, p. 150, emphasis added). Thus, the pendulum has been swinging back toward a more active media audience, which is, to some extent, reflective of the preindustrialized era of media-audience relations.[3]

The institutions that dominated the mass media era have been characterized as having an innate preference for the unidirectional model, in which audiences are more passive than active and more content consumers than content producers/distributors (Napoli, 2011). According to Shimpach (2005), "The ideal audience to emerge from the culture industry's construction is largely passive, observing the products of the culture industry, waiting around to be counted, measured, and receive intervention" (p. 350). Similarly, Angus (1994) argued that the tendency of most significant contemporary communications systems has been to produce audiences without the capacity to produce and circulate social knowledge.

Efforts to maintain this manifestation of the audience have taken a variety of forms (for a more detailed discussion, see Napoli, 2011). According to Benkler (2000), efforts at maintaining the traditional media consumer-producer relationship have been directed at three layers of the new media environment: the content layer (involving issues such as copyright and fair use), the logical layer (involving encryption), and the infrastructure layer (involving the structure and operation of communications networks). The infrastructure layer is of particular relevance to our analysis, given the concern with the differences between mobile and PC-based forms of Internet access.

On this front, an example such as the persistence of the asymmetrical bandwidth phenomenon—in which Internet service provider subscribers are often able to download content at much faster speeds than they can upload content (Winseck, 2002)—can be interpreted as an effort by media industry stakeholders to preserve the content provider-audience distinction that was much more clearly defined within the traditional media. This practice has been criticized as an effort to "turn users into simple appendages of the network" (Winseck, 2002, p. 807). One industry analyst described this situation as "a hangover of the old mass media days" (Jesdanun, 2006, p. 1).

This persistence of asymmetrical bandwidth is indicative of what Peter Lunenfeld (2011) has described as a broader, ongoing, "secret war between downloading and uploading" that has characterized the entire history of the computer and the Internet. These efforts are seldom completely successful. However, they can, as Yochai Benkler (2000) argued, raise "the costs of becoming a user—rather than a consumer—of information and undermine the possibility of becoming a producer/user of information" (pp. 562–563).

These are important points to keep in mind as we next consider the possible implications that the mobile conversion might have in relation to these broader institutional processes (Sawhney, 2009). However, the goal here is not to assert intentionality in relation to the mobile conversion. We are not arguing that the migration to mobile forms of Internet access is being orchestrated in a concerted effort to reassert the traditional content provider-audience dynamic—only that, regardless of the underlying rationales, this is exactly the effect it is having.

The Mobile Conversion and the
Repassification of the Media Audience

As previously noted, the process of media and audience evolution has been one in which the combination of technological and institutional factors has, in recent years, facilitated a return to a more active audience, more engaged in content creation than was the case in the mass media era. However, we contend that the migration of Internet access and usage from PC-based to mobile platforms[4] may facilitate another reversal—a return to the more passive, consumption-oriented audiences of the mass media era—at least relative to what was facilitated by the PC-based Internet access model. That is, the systemic migration to the mobile Internet may ultimately serve to increase the passivity of Internet audiences in the same way that the rise of broadcast media had brought greater passivity to the role of the audience in the media ecosystem (e.g., Angus, 1994; Butsch, 2008). In these ways, the systemic prioritization of the mobile Internet model—in terms of content creation, platform design, and access and usage—can be seen as a troubling evolutionary regression for the medium as a whole.

This argument is premised upon a continuation of the diffusion and usage patterns we see now, in which mobile Internet access is eclipsing PC-based access, and in which an increasing percentage of the population is mobile only and/or mobile native in its Internet usage. Presumably, therefore, an increasing proportion of the content and platforms available online may be designed and function in ways that prioritize mobile over PC-based access—something that is already starting to occur (Benton, 2014). As mobile users become the majority rather than the minority, it stands to reason that the online environment will be increasingly calibrated to serve them and to reflect the capabilities of the devices they use. Assuming that this becomes the case, we are likely in the midst of the development of a mobile-oriented Internet ecosystem that is in some ways an evolutionary step backward from the PC-based Internet ecosystem, across a number of important dimensions that bear directly on issues of audience activity and the extent to which users create and disseminate content. In the following sections, we explore factors such as differences in platform architecture and usage behaviors that contribute to this regressive dimension of the ongoing mobile conversion (for a broader discussion of the mobile Internet's deficiencies relative to PC-based forms of access, see Napoli & Obar, 2014).

Platform Architecture

One of the most important (and obvious) ways that mobile-based forms of Internet access represent an evolutionary regression relative to PC-based forms of access involves the nature of the devices themselves. Mobile handheld devices (as well as tablets) are a much less open platform for engaging with the Internet than the PC (for detailed discussions of this issue, see Wu, 2007; Zittrain, 2008). Horner (2011) noted that, "unlike personal computers (PC), mobile handsets are primarily closed, proprietary technologies that are difficult for people to adapt and programme for different uses" (p. 13).

As the late Steve Jobs stated in the days leading up to the launch of the iPhone, "These are more like iPods than they are like computers" (quoted in Sawhney, 2009, p. 106). In this regard, Jobs characterized mobile devices as essentially a regression to an older, unidirectional media consumption device (the iPod) rather than as something remodeling the capabilities of a computer. To the extent that these devices are being used instead of PCs or laptops, one could argue that they are being used to try to fulfill functions that they were never really designed to fulfill.

Users who go online via more closed, less programmable devices such as tablets and smartphones do not have anything close to the same capacity to enhance the Internet and its offerings in the same way or to reap the benefits of doing so. From this standpoint, mobile device connectivity to the Internet is, to some extent, contradictory to the very nature of the Internet's origins and growth, in which the opportunities to program and innovate enjoyed by the network's end users were key engines for the Internet's development (Zittrain, 2008). The more passified users of mobile devices have far less capacity for programming and innovation.

Usage Patterns

The differences in usage patterns that research has uncovered between mobile and PC-based contexts help to illustrate this point. Here, the focus is on two aspects of Internet usage: information seeking and content creation.

Information Seeking. One of the defining aspects of the Internet has been the way it facilitates and encourages active information seeking on the part of individuals, in contrast to the relatively more passive content reception dynamic that characterized legacy media. This shift is a function of the exponentially greater amount of content and sources that are accessible online compared to traditional media, as well as the associated tools for information seeking that can be utilized online. From this standpoint, the continuum of audience activity is partially a function of the extent to which individuals use a platform to actively engage in information seeking.

Comparative analyses of information seeking via PCs and mobile devices reveal some stark differences. Humphreys, Von Pape, and Karnowski (2013) conducted in-depth interviews with U.S. and German Internet users, characterizing PC-based Internet usage as primarily immersive and mobile Internet usage as primarily extractive (i.e., purposeful engagements of a shorter duration). Similarly, Isomursu, Hinman, Isomursu, and Spasojeciv (2007), who examined the dominant metaphors for the mobile Internet that emerged in a series of user studies, concluded that an appropriate metaphor for PC-based Internet access is scuba diving, in which individuals can "dive deep into their areas of interest and be totally immersed with the experience" (p. 262). Mobile Internet access, on the other hand, is analogous to snorkeling, because "Environmental factors and equipment are optimized for 'skimming the surface' or 'dipping in and out'" (Isomursu et al., 2007, p. 262). The authors concluded that "*Passive forms of content consumption . . .* often work better in this kind of situation because they take up less cognitive energy" (Isomursu et al., 2007, pp. 262–263, emphasis in original; see also Nielsen & Fjuk, 2010).

Supporting this perspective is a growing body of comparative research examining the dynamics of searching and information seeking. A usage study of mobile users in six countries concluded that information gathering was not a common task amongst mobile device users (Cui & Roto, 2008). A more recent study comparing user behavior across tablets and smartphones found that users viewed 70% more pages per website visit when using a tablet than when using a smartphone (O'Malley, 2013).

Along related lines, research suggests that the average number of characters in mobile search queries is significantly lower than the average number of characters in PC-based search queries. Mobile searches also utilize a significantly more limited search vocabulary than PC-based searches (Baeza-Yates, Dupret, & Velasco, 2007; Church, Smyth, Cotter, & Bradley, 2007). Mobile searches also are significantly less likely to utilize advanced search features such as Boolean operators or query modifiers (Church et al., 2007).[5] Mobile searchers exhibit a significantly greater tendency to rely on the first few search returns than PC-based searchers (Church, Smyth, Cotter, & Bradley, 2008). Thus, ranking effects (the extent to which placement in search returns affects content selection) have been shown to be significantly more powerful in mobile than in PC search contexts (Ghose, Goldfarb, & Han, 2013). Some of this research is obviously a bit dated by Internet standards, but it is important to keep in mind that many of the mobile natives that are coming online are doing so in developing countries and with devices that do not possess the most recent features and capabilities.

Together, this accumulation of findings illustrates how the mobile usage context compels users into a fundamentally more passive orientation, in which they engage in less active, sophisticated, and autonomous information seeking than is the case in the PC usage context.

Content Creation Versus Consumption. In the realm of content creation, there are a number of ways in which mobile access represents a step backward relative to PC-based access. Despite celebrated examples of bestselling novels being written on smartphones (Onishi, 2008), it is still the case that entering significant amounts of information is easier to accomplish on a PC-sized keyboard than on a mobile device's keypad (Yesilada, Harper, Chen, & Trewin, 2010).

Such differences ultimately cast the mobile device as more of an information retrieval device and less of an information creation and dissemination device than the PC. This is not to say that substantial amounts of content cannot be created and distributed via mobile devices (e.g., VozMob Project, 2011)—only that the creation of content of significant scope, complexity, and depth is much more easily accomplished via PCs than it is via mobile devices. Certainly, if we follow this line of thinking into the realm of complex, large-scale applications and services, the gap between the two modes of access likely can never be overcome. Consider, for instance, whether a Facebook or a Google could be created on a smartphone. The ability of the end user to make significant contributions to the Internet, in terms of content or applications, is significantly curtailed relative to what can be achieved via a PC if the end user is limited to a mobile device.

This perspective is supported by findings that the creation of large and complex documents is an uncommon activity on mobile devices (Cui & Roto, 2008; Yesilada et al., 2010). The Wikimedia Foundation has found that Wikipedia participants who transition from PCs/laptops to tablets contribute significantly less after the transition, because tablets "are better for watching videos and surfing the Internet than for typing text" (Walker, 2013, p. 1). It would seem reasonable, then, to expect an even more pronounced disparity between PC/laptop-based and smartphone-based contributors, given the more limited interface capabilities of mobile devices relative to tablets.

Kaikkonen (2011) compared mobile versus PC findings across two time periods (2007 and 2010) and found that a number of significant differences persist, particularly in terms of production- and dissemination-oriented activities such as writing e-mail and participating in online discussions. For instance, in 2010, 35% of mobile subjects reported writing e-mails on their devices, compared with 57% of PC users. Similarly, 24% of mobile users reported participating in online discussions, compared with 51% of PC users (Kaikkonen, 2011). Similar patterns were found in a PC deprivation study in which PC users were allowed only mobile Internet access for four days (Hinman, Spasojevic, & Isomursu, 2008). Participants used the device to access their e-mail but significantly cut back on composing and sending e-mails due to the difficulties associated with composing e-mails on a mobile device (Hinman et al., 2008; see also Bao, Pierce, Whittaker, & Zhai, 2011).[6]

Another recent study, which focused on South Korea—a country with high levels of Internet and mobile penetration—found significant interactions between content creation and content consumption activities (Ghose & Han, 2011). Specifically, through time-lagged analysis of data on content uploading and downloading activities (obtained from nearly 200,000 3G mobile users), the authors found a negative causal relationship between the two. In effect, content consumption activities appear to be displacing content creation activities to a significant extent.

These studies' findings appear to be reflective of broader patterns that have emerged in more macrolevel, multinational research that indicates that "as new users get online, fewer and fewer of them appear to be content producers" (Pimienta, 2008, p. 31). Although this pattern may reflect intrinsic differences between earlier and later Internet adopters, and it may diminish over time, it also seems reasonable to ask whether such patterns might be a reflection of the differences in the characteristics of the devices they are using to access the Internet. Perhaps the extent to which mobile devices are less conducive to substantive content creation and dissemination is contributing to the fact that more recent Internet adopters (who are predominantly mobile users) are producing less content than previous waves of Internet adopters.

Further, as was noted previously, many mobile-only users are accessing the Internet for the first time via mobile devices. This stands in stark contrast to the earlier pattern, in which an individual first accessed the Internet via the PC and later migrated to the mobile device (either exclusively or in conjunction with the PC). The difference in these processes has import, because the PC-initiated Internet user may have developed certain types of knowledge and skills that transfer to the mobile context, thereby allowing a PC-initiated user of a mobile device to make more effective use of the platform than a mobile native (Hyde-Clark & Van Tonder, 2011). For instance, a multinational study of mobile users in developing nations found that the single best predictor of usage of Internet services on smartphones is whether the individual user already used the Internet through a PC (Zainudeen &

Ratnadiwakara, 2011). A more recent study, of young people in Sri Lanka, found that underprivileged youths utilized a much narrower range of their smartphones' capabilities (in many cases not moving beyond traditional voice functionality) than did those of higher socioeconomic status, due largely to a lack of relevant competencies (Wijetunga, 2014).

Hargittai and Kim (2010) found consistent evidence that the amount of prior Internet experience and the range of Internet-related skill sets that individuals have developed in the context of PC-based Internet access are positively related to the range of functionalities that these individuals utilize on their mobile devices. On the basis of these findings, they raise the concern that the spread of mobile devices could exacerbate existing digital inequalities and feed into increasing disparities, in what John Horrigan termed "digital readiness"—the knowledge and skill sets necessary to use the Internet effectively (quoted in Stewart, 2013, online). From the standpoint of our analysis, the key implication is that the next generation of Internet users seems likely to be significantly underskilled relative to the previous users and therefore less equipped to engage in the same degree of active, sophisticated information seeking and independent content creation and dissemination as the previous generation.

The Mobile Conversion and the Suppression of the Internet's Radical Potential

We have taken an admittedly one-sided perspective in our exploration of the implications of the mobile conversion, as a counterpoint to the overwhelmingly positive discourse accompanying the diffusion of mobile-based forms of Internet access thus far. The situation is certainly more complex and multidimensional than it has been characterized here. However, our point is that there do appear to be significant dimensions across which the ongoing mobile conversion of the Internet represents an evolutionary regression of the online audience, one in which audiences regress on the active-passive continuum, and one in which the distinction between content creators and content consumers that had become increasingly blurred over the past two decades once again becomes more distinct.

These patterns represent a significant diminishment of some of the Internet's most important capacities; and, when looked upon in hindsight, they may fit quite well into the larger narrative of the "suppression of radical potential" (Winston, 1986, p. 18) that has characterized much of the history of media. The "suppression of radical potential" refers to the consistency with which institutional forces have effectively limited the extent to which new media technologies can realize their full potential to disrupt established conditions (Winston, 1986). In this case, it appears that the mobile conversion is, in some ways, suppressing the radical potential of the Internet.

From this standpoint, it is important that we try to acknowledge the potentially detrimental significance of the current trajectory in the evolution of the Internet and its users. As Lunenfeld (2011) argued, "If the recent Web, linked to desktops and fully featured laptops becomes an anomaly on the path to an ever more one-sided consumer mobility . . . we will have made a major mistake" (p. 82).

Notes

1. The term *passification* is used here as something separate and distinct from the more familiar term "pacification." To pacify means to make an individual or group calm and/or peaceful. Here, the focus is on making individuals or groups passive—as in not active but instead inclined to be acted upon, thus the use of the term "passification" for the purposes of this analysis.

2. We use the term "PC" throughout this chapter in reference to both desktop- and laptop-based forms of Internet access.

3. This pattern can be seen as something of a corollary to Wu's (2011) observation regarding the historical evolution from open to closed media systems.

4. This analysis employs a dichotomy between PC-based and mobile device-based forms of Internet access. For the purposes of our analysis, "mobile devices" refer primarily to smartphones. Tablets, which in many ways seem to represent an intermediate position between PCs/laptops and smartphones, are, for the purposes of this analysis, not being considered as a significant component of the mobile conversion, due largely to the fact that the bulk of tablet usage actually adheres to the dynamics of PC/laptop usage, in terms of taking place primarily in the home via a residential broadband subscription. A very small percentage of tablets operate with the kind of mobile data plans characteristic of smartphones (for a more detailed discussion of the dynamics of tablet usage, see Napoli & Obar, 2014). However, in a few instances, this analysis refers to studies of tablet usage, in cases in which the findings seem to have implications for the PC/laptop-mobile device dichotomy being discussed here.

5. These results come from an analysis of mobile usage data gathered from over 600,000 European mobile Internet users.

6. The small scale of this study, which focused on eight U.S. college students (data derived from online diaries that they kept of their activities via their mobile devices), should be noted.

References

Angus, I. (1994). Democracy and the constitution of audiences: A comparative media theory perspective. In J. Cruz & J. Lewis, (Eds.), *Viewing, reading, listening: Audiences and cultural reception* (pp. 233–252). Boulder, CO: Westview.

Arthur, C. (2011, June 5). How the smartphone is killing the PC. *The Guardian.* Retrieved from http://www.guardian.co.uk/technology/2011/jun/05/smartphones-killing-pc

Baeza-Yates, R., Dupret, G., & Velasco, J. (2007, May). *A study of mobile search queries in Japan.* Paper presented at the WWW2007 Conference, Banff, Canada. Retrieved December 16, 2014 from http://www2007.org/workshops/paper_50.pdf

Bao, P., Pierce, J., Whittaker, S., & Zhai, S. (2011). *Smart phone use by non-mobile business users.* Paper presented at the MobileHCI Conference, Stockholm, Sweden.

Benkler, Y. (2000). From consumers to users: Shifting the deeper structures of regulation toward sustainable commons and user access. *Federal Communications Law Journal, 52*(3), 561–579.

Benkler, Y. (2006). *The wealth of networks: How social production transforms markets and freedom.* New Haven, CT: Yale University Press.

Benkler, Y., Roberts, H., Faris, R., Solow-Niederman, A., & Etling, B. (2013). *Social mobilization and the networked public sphere: Mapping the SOPA-PIPA debate* (SSRN Scholarly Paper No. ID 2295953). Rochester, NY: Social Science Research Network.

Benton, J. (2014, July 18). In the (appropriate!) shift to mobile, is desktop sometimes being abandoned too quickly? *Nieman Journalism Lab.* Retrieved December 16, 2014 from http://www.niemanlab.org/2014/07/in-the-appropriate-shift-to-mobile-is-desktop-sometimes-being-abandoned-too-quickly/

Billings, V. (1986). Culture by the millions: Audience as innovator. In S. J. Ball Rokeach & M. G. Cantor (Eds.), *Media, audience, and social structure* (pp. 206–213). Beverly Hills, CA: Sage.

Bold, W., & Davidson, W. (2012). Mobile broadband: Redefining internet access and empowering individuals. In S. Dutta & B. Bilbao-Osorio (Eds.), *The global information technology report* (pp. 67–77). Geneva, Switzerland: World Economic Forum and INSEAD. Retrieved December 16, 2014 from http://academy.itu.int/moodle/pluginfile.php/38548/mod_resource/content/1/GITR_2012.pdf

Butsch, R. (2008). *The citizen audience: Crowds, publics, and individuals.* New York, NY: Routledge.

Castells, M., Fernandez-Ardevol, M., Qiu, J. L., & Sey, A. (2007). *Mobile communication and society: A global perspective.* Cambridge, MA: The MIT Press.

Castronova, E. (2001). Virtual worlds: A first-hand account of market and society on the cyberian frontier (CESifo Working Paper Series No. 618). Retrieved December 16, 2014 from http://ssrn.com/abstract=294828

Chigona, W., Beukes, D., Vally, J., & Tanner, M. (2009). Can mobile Internet help alleviate social exclusion in developing countries? *Electronic Journal on Information Systems in Developing Countries, 36*(7), 1–16.

Chircu, A. M., & Mahajan, V. (2009). Revisiting the digital divide: An analysis of mobile technology depth and service breadth in the BRIC countries. *Journal of Product Innovation Management, 26,* 455–466.

Church, K., Smyth, B., Cotter, P., & Bradley, K. (2007). Mobile information access: A study of emerging search behavior on the mobile Internet. *ACM Transactions on the Web, 1*(1), 1–38.

Church, K., Smyth, B., Cotter, P., & Bradley, K. (2008). A large scale study of European mobile search behavior. *Proceedings of the 2008 Mobile HCI Conference* (pp. 13–22). New York, NY: ACM.

Cover, R. (2004). New media theory: Electronic games, democracy and reconfiguring the author-audience relationship. *Social Semiotics, 14*(2), 173–191.

Cover, R. (2006). Audience inter/active: Interactive media, narrative control and reconceiving audience history. *New Media & Society, 8*(1), 139–158.

Cui, Y., & Roto, V. (2008). *How people use the web on mobile devices.* Paper presented at the International World Wide Web Conference, Beijing, China.

Dutton, W. H. (2009). The fifth estate emerging through the network of networks. *Prometheus, 27*(1), 1–15.

Fong, M. W. (2009). Technology leapfrogging for developing countries. *Electronic Journal of Information Systems in Developing Countries, 36*(6), 1–12.

Ghose, A., Goldfarb, A., & Han, S. P. (2013). How is the mobile internet different? Search costs and local activities. *Information Systems Research, 24*(3), 613–631.

Ghose, A., & Han, S. P. (2011) An empirical analysis of user content generation and usage behavior on the mobile internet. *Management Science, 57*(9), 1671–1991.

Griffen-Foley, B. (2004). From Tit-Bits to Big Brother: A century of audience participation in the media. *Media, Culture, and Society, 26*(4), 533–548.

Hargittai, E., & Kim, S. J. (2010). *The prevalence of smartphone use among a wired group of young adults* (Working Paper). Evanston, IL: Institute for Policy Research, Northwestern University. Retrieved December 16, 2014 from http://www.ipr.northwestern.edu/publications/docs/workingpapers/2011/IPR-WP-11-01.pdf

Harrison, T. M., & Barthel, B. (2009). Wielding new media in Web 2.0: Exploring the history of engagement with the collaborative construction of media products. *New Media and Society, 11*(1/2), 155–178.

Hinman, R., Spasojevic, M., & Isomursu, P. (2008). *They call it "surfing" for a reason: Identifying mobile internet needs through PC deprivation.* Paper presented at the Computer Human Interaction Conference, Florence, Italy.

Horner, L. (2011). A human rights approach to the mobile Internet. *Association for Progressive Communications.* Retrieved December 16, 2014 from https://www.apc.org/en/system/files/LisaHorner_MobileInternet-ONLINE.pdf

Horrigan, J. B. (2009). *Wireless internet use.* Washington, DC: Pew Internet and American Life Project.

Humphreys, L., Von Pape, T., & Karnowski, V. (2013). Evolving mobile media use: Uses and conceptualizations of the mobile Internet. *Journal of Computer-Mediated Communication, 18,* 491–507.

Hyde-Clark, N., & Van Tonder, T. (2011). Revisiting the leapfrog debate in light of current trends of mobile phone usage in the greater Johannesburg area, South Africa. *Journal of African Media Studies, 3*(2), 263–276.

Ishii, K. (2004). Internet use via mobile phone in Japan. *Telecommunications Policy, 28,* 43–58.

Isomursu, P., Hinman, R., Isomursu, M., & Spasojeciv, M. (2007). Metaphors for the mobile internet. *Knowledge, Technology & Policy, 20*(4), 259–268.

Jesdanun, A. (2006, December 18). Imbalance in net speeds impedes sharing. *The Washington Post.* Retrieved Sept. 18, 2008 from http://www.washingtonpost.com/wp-dyn/content/article/2006/12/18/AR2006121800610.html

Kaikkonen, A. (2011). Mobile internet, internet on mobiles or just internet you access with variety of devices? *Proceedings of the 23rd Australian Computer-Human Interaction Conference* (pp. 173–176). New York, NY: Association for Computing Machinery.

Lastowka, G. (2010). *Virtual justice.* New Haven, CT: Yale University Press.

Livingstone, S. (2003). The changing nature of audiences: From the mass audience to the interactive media user. In A. Valdivia (Ed.), *Companion to media studies* (pp. 337–359). Oxford, UK: Blackwell.

Lunenfeld, P. (2011). *The secret war between downloading and uploading: Tales of the computer as culture machine.* Cambridge, MA: The MIT Press.

Napoli, P. M. (2011). *Audience evolution: New technologies and the transformation of media audiences.* New York, NY: Columbia University Press.

Napoli, P. M., & Obar, J. A. (2014). The emerging internet underclass: A critique of mobile internet access. *The Information Society, 30,* 223–234.

Newman, K. M. (2004). *Radio active: Advertising and consumer activism, 1935–1947.* Berkeley: University of California Press.

Nielsen, P., & Fjuk, A. (2010). The reality beyond the hype: Mobile internet is primarily an extension of PC-based internet. *The Information Society, 26*(5), 375–382.

Obar, J. A. (2014). Canadian Advocacy 2.0: An analysis of social media adoption and perceived affordances by advocacy groups looking to advance activism in Canada. *Canadian Journal of Communication, 39*(2), 211–233. Retrieved December 16, 2014 from http://ssrn.com/abstract=2254742

Obar, J. A., & Shade, L. R. (2014). *Activating the fifth estate: Bill C-30 and the digitally-mediated public watchdog.* Retrieved from http://ssrn.com/abstract=2470671

Obar, J. A., Zube, P., & Lampe, C. (2012). Advocacy 2.0: An analysis of how advocacy groups in the United States perceive and use social media as tools for facilitating civic engagement and collective action. *Journal of Information Policy, 2,* 1–25. Retrieved December 16, 2014 from http://ssrn.com/abstract=1956352

O'Malley, G. (2013, November 19). Consumers consider phones primary mobile device. *MediaPost.* Retrieved December 16, 2004 from http://www.mediapost.com/publications/article/213847/consumers-consider-phones-primary-mobile-device.html

Onishi, N. (2008, January 20). Thumbs race as Japan's bestsellers go cellular. *The New York Times.* Retrieved December 16, 2014 from http://www.nytimes.com/2008/01/20/world/asia/20japan.html?pagewanted=all&_r=0

Pimienta, D. (2008). Accessing content. In A. Finlay & L. Nordstrom (Eds.), *Global information society watch* (pp. 31–33). Melville, South Africa: Alliance for Progressive Communications, Hivos, and ITeM. Retrieved December 16, 2014 from https://www.apc.org/en/system/files/GISW2008_EN.pdf

Purcell, K., Rainie, L., Rosenstiel, T., & Mitchell, A. (2011). *How mobile devices are changing community information environments.* Washington, DC: Pew Project for Excellence in Journalism.

Razlogova, E. (1995). *The voice of the listener: Americans and the radio industry, 1920–1950* (Unpublished doctoral dissertation). George Mason University, Fairfax, VA.

Ross, S. J. (1999). The revolt of the audience: Reconsidering audiences and reception during the silent era. In M. Stokes & R. Maltby (Eds.), *American movie audiences: From the turn of the century to the early sound era* (pp. 92–111). London, UK: British Film Institute.

Sawhney, H. (2009). Innovation at the edge: The impact of mobile technologies on the character of the internet. In G. Goggin & L. Hjorth (Eds.), *Mobile technologies: From telecommunications to media* (pp. 105–117). New York, NY: Routledge.

Schejter, A., & Tirosh, N. (2012). Social media new and old in the Al-'Aarakeeb conflict: A case study. *The Information Society, 28*(5), 304–315.

Shimpach, S. (2005). Working watching: The creative and cultural labor of the media audience. *Social Semiotics, 15*(3), 343–360.

Stewart, A. (2013, November 8). Horrigan: Stop using "digital divide" phrase. *Techwire.* Retrieved December 16, 2014 from http://techwire.net/horrigan-stop-using-digital-divide-phrase/

Svoen, B. (2007). Consumers, participants, and creators: Young people's diverse use of television and new media. *ACM Computers in Entertainment, 5*(2), 1–16.

VozMob Project. (2011). Mobile voices: Projecting the voices of immigrant workers by appropriating mobile phones for popular communication. In P. M. Napoli & M. Aslama (Eds.), *Communications research in action: Scholar-activist collaborations for a democratic public sphere* (pp. 177–196). New York, NY: Fordham University Press.

Walker, R. (2013, March 4). Are smartphones and tablets turning us into sissies? *Yahoo! News.* Retrieved December 16, 2014 from http://news.yahoo.com/are-smartphones-and-tablets-turning-us-into-sissies--175359859.html

Wareham, J., Levy, A., & Shi, W. (2004). Wireless diffusion and mobile computing: Implications for the digital divide. *Telecommunications Policy, 28,* 439–457.

Wijetunga, D. (2014). The digital divide objectified in design: Use of the mobile telephone by underprivileged youth in Sri Lanka. *Journal of Computer-Mediated Communication, 19*(3), 712–726.

Winseck, D. (2002). Netscapes of power: Convergence, consolidation, and power in the Canadian mediascape. *Media, Culture, and Society, 24,* 795–819.

Winston, B. (1986). *Misunderstanding media.* Cambridge, MA: Harvard University Press.

Wu, T. (2007). Wireless Carterfone. *International Journal of Communication, 1,* 389–426.

Wu, T. (2011). *The master switch: The rise and fall of information empires.* New York, NY: Knopf.

Yesilada, Y., Harper, S., Chen, T., & Trewin, S. (2010). Small-device users situationally impaired by input. *Computers in Human Behavior, 26,* 427–435.

Zainudeen, A., & Ratnadiwakara, D. (2011). Are the poor stuck in voice? Conditions for adoption of more-than-voice mobile services. *Information Technologies & International Development, 7*(3), 45–59.

Zittrain, J. (2008). *The future of the Internet: And how to stop it.* New Haven, CT: Yale University Press.

Social Media Audience Metrics as a New Form of TV Audience Measurement

Darryl Woodford, Ben Goldsmith, and Axel Bruns

The Uncertain Business of Audience Measurement

Understanding and acting on the behavior of media audiences is a multibillion-dollar business. Broadcasters and other media providers, advertisers, advertising agencies, media planners, and audience research companies have significant financial stakes in the collection and analysis of audience data. In addition, policy makers, academics, and audience members themselves have interests in the technologies and methodologies used to measure audiences, as well as in the data and their uses. But the audience rating convention—the necessary consensus among stakeholders about who and what is counted, how the counting is done, how the data are interpreted and valued—is under pressure as never before. Digitization, media convergence, and audience fragmentation have dramatically disrupted the business of audience measurement. New metrics and analytical systems have been developed to answer some of the questions raised by technological change, but they are also posing challenges to stakeholders about their capacity to deal with the explosion of raw and customized data on audience behavior. The volume of information that is available for aggregation and analysis has grown enormously, but with that growth has come a host of uncertainties about audience measurement, and in particular, about the broadcast ratings system.

Uncertainty has driven an extraordinary research effort, a flight to accountability, in which a proliferating number of information and research companies have tried to make sense of the accumulating data about media use, often with conflicting results. This was one of the reasons behind what Alan Wurtzel (2009), president of research at American broadcaster NBC, has called the "crisis in measurement," although the wealth of data and the efforts being made to analyze that data may mean that this period could be looked back on as a golden age if the industry's ideal scenario—the collection, cross-tabulation, and fusing of massive amounts of data and large datasets—can be realized. This would potentially produce the advertising industry's holy grail: single source, or consumer-centric holistic measurement (World Federation of Advertisers, 2008), although serious questions would also arise, not least about privacy and audience members' and consumers' awareness of the data being collected (Andrejevic, 2007, 2013).

In some respects, the current state of uncertainty is nothing new. Historically, the introduction or expansion of commercial broadcasting services; changes to the structure, economics, technologies, or the policy field of broadcasting; and evolving patterns of audience behavior have all spurred the development of new technologies, methodologies, and rationales for quantifying television audiences. For various reasons primarily to do with establishing the parameters for the buying and selling of airtime in predominantly commercial or mixed public service/commercial broadcasting markets in countries around the world, consensus has tended to form around the need for an authoritative, simple measure of exposure—who is watching television, which channel or service are they watching, and for how long. There has long been great (and recently, increasing) interest in measuring audience members' engagement with programming and advertising—how much attention they are paying, what their opinion is about what they are watching, and what impact the program or commercial has on them—but exposure has remained the standard for measuring broadcast ratings and the core of the ratings convention ever since Archibald Crossley's first survey of American radio listeners in 1929 (Balnaves, O'Regan, & Goldsmith, 2011). Despite the contemporary crisis, which is multifaceted, ratings data are still and will continue to be in demand because there will always be the need for common currencies for buying and selling advertising and program content. There will undoubtedly be changes in the practicalities of audience measurement, particularly given the challenges presented by the likely spread of broadband-enabled set-top boxes, which have been described by the Council for Research Excellence (2010) as the "wild west" (p. 4) of research.

The availability of multiple channels through subscription or free-to-air television, coupled with ever-increasing online video options, amplifies viewer/consumer choice and consequently distribute the available audience much more widely than earlier broadcasting systems. Napoli (2003, p. 140) argued that this fragmentation increases the disparity between the predicted and the measured audience and reduces the reliability of data collected in traditional sample-based methods. It certainly increases what has been called the "research ask" and complicates the carefully calibrated equations that produce the ratings. Although mass audiences can still be assured for certain major events, often live international sports championships, audiences in general have dispersed. Content providers, advertisers, and research organizations have had to track not only time-shifting and catch-up TV but also migration across platforms and even beyond the home. Audience fragmentation has precipitated proliferations of data, methods, metrics, and technologies that in turn have allowed samples of a few hundred, in panels or diaries, to multiply into surveys of millions of subscribers and produce competing currencies. Opportunities for advertisers to reach consumers through media and other touchpoints have proliferated, while advertisers' and content providers' desire for solid numbers and discontent with the prevailing currency and methods have opened spaces for research and analysis.

Public service broadcasters have typically been more interested than their commercial counterparts in qualitative research that provides detailed information about audience enjoyment and engagement with programming. This, for example, was the focus of audience research conducted by and for the BBC from the mid-1930s (Silvey, 1951). For commercial broadcasters—as, eventually, for public service broadcasters, too—ratings have served a range of purposes, from measuring the popularity of particular programs, providing guidance in program planning and scheduling, informing service delivery, keeping abreast of change in audience tastes and practices, and establishing the value of time sold for advertising. Ratings can also act as a proxy for the broadcaster's share price and an indicator of (and influence upon) its overall financial health (Balnaves et al., 2011). For advertising agencies, media planners, and advertisers themselves, ratings help determine how much will be spent on advertising on a particular channel or network, as well as where and when advertisements will be placed. But ratings are not only used within broadcasting. They are also of interest to the mainstream media and the public at large for what they appear to reveal about the success or otherwise of programs and broadcasters, as well as to academics, media critics, and public authorities who "use, quote, debate and question" ratings (Bourdon & Méadel, 2014, p. 1). In Canada, for example, the media regulator uses ratings as one measure to judge the success of CanCon (Canadian Content) drama policies and as the basis on which funding for future production is allocated (Savage & Sévigny, 2014). Criticism of the ratings has come from many quarters and taken many forms, from theoretical and technical questioning of the methodologies and technologies deployed over time to concerns about the business practices of data suppliers and the tendency of those who use ratings to "endow the audience with a reality and thereness it does not possess" (Balnaves et al., 2011, p. 229). Yet, despite the disruption wrought by digitization, a variety of parties continue to maintain a variety of interests in the collection of robust, reliable, and commonly agreed-upon metrics about audiences, as well as in agreeing on what counts as an audience. Alan Wurtzel observed the situation in the United States in 2009:

> A couple of years ago, Nielsen delivered a single TV-rating data stream. Today, Nielsen routinely delivers more than two dozen streams (yes, we counted them) and countless more are available for any client willing to pull the data. Moreover, set-top boxes (STB), moving closer and closer to second-by-second data, will produce a staggering amount of new information. And, with internet and mobile metrics as well, it's not the amount of data that is the problem; it's the quality and utility. (p. 263)

This is the key challenge for ratings providers in the future: providing quality and useful data. But given that so much is in flux, including common understandings of "quality" and "utility," it appears for the moment as though multiplication of research vehicles and partnerships will inevitably continue as ratings companies jostle over currencies and simultaneously provide bespoke services to individual clients.

In addition to Nielsen's multiple streams and the wealth of other services available, broadcasters, content providers, and advertisers must also contend with the power of bottom-up systems of recommendation and rating that have emerged with the Internet. From Facebook's "Like" option, which allows readers to signify in a single click their approval or appreciation of something posted by a friend (importantly, there is no option to "Dislike"), through sharing and retweeting on Twitter, to supporting (or Digg-ing) something posted to Digg.com, the opportunities for audiences to register opinions or rate all sorts of things on the Internet are many and varied. To varying degrees, research companies, advertisers, and content providers are realizing the importance of social media in gauging audience opinions about the quality of content. The characteristic online behavior of countless people now routinely involves what futurist Mark Pesce (2006) called "the three Fs:" finding, filtering, and forwarding information found online to contacts (or followers in Twitter-speak, friends on Facebook). In contrast to more restricted media such as free-to-air broadcast television, audiences can now find desired audiovisual content, or close approximations, online. The actions of tagging, rating, and recommending function as forms of feedback, often for the principal benefit of the audience's own network. But ever more sophisticated and insistent forms of monitoring behavior and turning it into useful data are capturing this information, adding it to databases for dissection and fusion. In terms of quality, audience members who follow or forward content on multiple media are exactly the audiences that producers of media content are trying to cultivate, in part because of the ratings they may provide in the future. It is the most committed, the most voracious of the online explorers or pioneers, the keenest edge of the community that evolves around content, who can shape the media choices of those around them, who will be most highly valued by producers, if not always by advertisers. All of these developments point to the likelihood that measures of popularity in social media will become more extensive in the future. In the remainder of this chapter, we discuss the ways in which particular forms of social media analysis can produce useful and actionable data about engagement with television that augment and extend the ratings' core focus on exposure.

Toward Social Media-Derived Audience Metrics

Traditional television ratings schemes provide a standardized and broadly reliable, but ultimately limited and one-sided, measure of audience interest; historically, they provide information on what audience research could readily and regularly quantify but fail to offer any fine-grained, in-depth evaluation of audience activities even from a quantitative perspective, much less from a qualitative one. In the emerging multichannel, multiplatform, multiscreen environment, they become manifestly insufficient.

Audiences for televisual content now access their shows through a range of channels: In addition to conventional reception of the live television broadcast, they may also utilize streaming cable or broadband catch-up services, time-shifted pause and rewind functionality, or (official or unauthorized) video downloads. Such services may be offered by a wide range of providers and platforms, including the original domestic broadcasters, their counterparts in other geographic regions (where new shows may screen ahead of the domestic broadcast date and become accessible to users outside the region through the use of geo-masking VPN services), by video streaming platforms such as YouTube (where content may have been uploaded by production companies, one or several regional broadcasters, or fans), and by download services from the Apple Store to BitTorrent file-sharing sites.

Audience engagement with such content remains identifiable and quantifiable in most of these cases: On-demand platforms from official catch-up services to unauthorized file-sharing sites generate their own usage metrics, even if they are not always shared publicly. To date, however, such metrics have yet to be aggregated and standardized in any reliable form; a number of scholarly and industry research projects have attempted to do so for individual platforms, but several such studies, especially by industry-affiliated market research organizations, are also flawed by an underlying agenda to promote fledgling on-demand services or prove the impact of content piracy.

Further, significant challenges exist in ascribing meaning to these metrics. Mere figures describing the number of requests for specific on-demand video streams or downloads may be misleading if they turn out to be inflated by multiple requests from the same user due to poor server performance or broadband throughput; even unique user figures may be misleading if there is a significant influx of audiences from outside the intended region of availability through the use of VPNs and other mechanisms. Recent research suggests, for example, that the streaming service Netflix has already gained a 27% share of the overall on-demand market in Australia, even though Netflix does not officially operate in that country (Ryall, 2014). Australian Netflix users' activities are therefore likely to inflate the metrics of the U.S. platform to which they have managed to connect.

Figures for on-demand requests and downloads also fail to accurately capture the quality of engagement with the televisual content thus accessed: Was a downloaded video actually watched? Did viewers watch the entirety of the program? Here, in spite of their own limitations, even conventional television ratings provide a more comprehensive picture of audience engagement, because they are at least able to track audience sizes at regular intervals during a broadcast and thus offer a glimpse of audience attrition or accretion rates. In their use of demographically representative panels of television households, such conventional ratings also continue to provide more detailed data on the popularity of specific programming with particular audience segments; this is likely to be absent from the metrics for alternative channels, where demographic data are often rudimentary at best.

Such information is especially crucial for broadcasters in the public service media sector, where an application of conventional ratings to programming, which is often deliberately designed to address specific niche interests and audiences, can significantly misjudge the ability of such programming to meet its intended aims. Here, evaluating the forms and quality of audience engagement is often more important than simply measuring the total size of the audience. But for commercial television channels, too, such information provides important clues that feed back into the design and production of new programming; there is, therefore, a significant need to move beyond the limitations of merely quantitative audience measurements.

Media, communication, and cultural studies scholarship has long recognized the active audience of mass media programming (Fiske, 1992) but has traditionally found it difficult to measure the extent and impact of audience activities or provide comprehensive qualitative evidence beyond individual small-scale case studies. That is, scholarship in this field has established the necessary conceptual tools for evaluating and understanding diverse forms of audience engagement but has so far lacked access to a substantial base of evidential data on audience activity to which such tools may be usefully applied to determine and categorize the forms of audience engagement with media content that are prevalent in the contemporary media ecology or to evaluate their meaning and relevance in the context of the specific public service and/or commercial aims pursued by media organizations.

This situation has shifted markedly in recent years, especially due to the emergence of second-screen engagement through social media as an audience practice that accompanies the viewing of televisual content. Such engagement has turned the active audience of television into a *measurably* active audience that generates a rich trail of publicly available evidence for its activities, and this trail can be gathered through the Application Programming Interfaces (APIs) of mainstream social media platforms or internally from the access logs of the engagement platforms operated by broadcasters themselves. With the computational turn (Berry, 2011) in humanities research, such data may now be used to test and verify the conceptual models for audience engagement that have been developed by media, communication, and cultural studies disciplines, both to quantify the level of such activity for individual broadcasters and their programming and to benchmark the quality of this engagement against the aims and ambitions set by the content producers.

This focus on using social media activities as an indicator of audience engagement is not without its own limitations, however. In the first place, social media audience metrics require active television audiences also to be active *on social media* and may thus privilege particular audience demographics that are especially likely to be using platforms such as Facebook and Twitter to discuss their television viewing. Further, social media-based engagement with televisual content is likely to be greatest when individual users can engage with other viewers of the same programming in close to real time; such metrics continue to privilege live or close to live viewing (through conventional broadcast or streaming services) rather than significantly time-shifted access. For major television events, a considerable social media audience around a shared televisual text is likely to persist at least for several hours, perhaps even days, before and especially after the live broadcast; the measurement of social media audience activities need not necessarily require exactly simultaneous engagement with the same text. This is demonstrated by the global social media response to television events such as new episodes of popular series from *Doctor Who* to *Game of Thrones,* which are typically screened in different time slots but in close temporal proximity to each other in different broadcast regions around the world.

If such limitations inherent in the data derived from television-related social media activities can be successfully negotiated, then a range of new opportunities for quantifying as well as qualifying audience engagement with televisual content emerge. First, a number of comparatively simple audience metrics may be established, including the volume of postings that relate to specific programming and the number of unique users generating such audience responses. Here, the substantially improved precision of public social media data compared to conventional ratings data makes it possible to identify almost to the second which moments in a particular broadcast generated the greatest audience response and thus how such activity ebbed and flowed with the progress of the show; a measurement of unique active users over the course of the broadcast also offers first insights into the influx or exodus of viewers. Various contextual factors must also be considered in such analysis, however—different program types and formats may lend themselves more or less well to continued social media activity, for example: Audiences may be glued to the screen during drama programming and post social media updates only during commercial breaks, while during political talk shows, they may be more prepared to respond to the panelists' statements on a continuous basis.

Additionally, publicly available background data derived from the social media platforms themselves may also be brought to bear on the analysis: For example, in addition to measuring the total number of users participating in a social media conversation about a given show, it would also be possible to determine the number of social media friends or followers for each user's account and thus to evaluate the extent to which the broadcast has been able to attract highly networked (which may be read as "influential") participants. Similarly, if background data exist not just about the size of such friendship networks but also about their structure (as Bruns, Burgess, & Highfield, 2014, have developed it for the Australian Twittersphere, for example), it becomes possible both to pinpoint the location of individual users within that network and to determine the total footprint of a particular program within the overall social media platform.

Such indicators begin not just to quantify total engagement but to provide a postdemographic alternative to the audience segmentation models of conventional ratings: Because social media networks are often structured not primarily according to geographic or sociodemographic factors but by similarities in interests, this approach to analyzing social media-based audience activities offers insights into whether a specific broadcast was able to achieve deep engagement with those sections of the overall network that are particularly concerned with the broadcast's topics and/or whether it managed to generate broad engagement irrespective of users' day-to-day interests and preferences (cf. Figure 9.1). Depending on broadcaster and program type, either or both of these objectives may be desirable: A political talk show on a niche public broadcast channel may seek deep engagement with a narrowly defined group of so-called political junkies (Coleman, 2003), whereas a broad-based entertainment show on a major commercial station would aim for responses from as broad a public as possible. Again, it should be noted that such analyses assume that engagement by the social media audience either provides a reasonable approximation of engagement by the wider television audience beyond specific social media platforms or that it is possible to correct for the demographic and postdemographic skews in the measurement of audience interests and activities that such a focus on social media-based engagement activities produces.

Figure 9.1. Social media footprints of different TV programming in Australia

Twitter-based engagement is shown with the political talk show *Q&A* (left) and the 2014 *Grand Final* of the Australian Football League (right). Against the backdrop of a follower network map for the 140,000 most connected Twitter accounts in Australia (in light gray), actively tweeting accounts for either broadcast are shown in black. *Q&A* tweeters are recruited predominantly from a network cluster focusing on politics (top left); *Grand Final* tweeters form a cluster focusing on sports (top center), but with much wider take-up across the Australian Twittersphere.

Finally, the immediate availability of audience members' social media responses to specific televisual programming also enables a qualitative analysis of their reactions beyond mere engagement metrics. It becomes possible, for example, to extract from the content of audience posts the key themes and topics of their responses, which may highlight the names of popular (or at least controversial) public figures, organizations, and actors, and to chart their relative centrality to the programming over the course of individual episodes or entire seasons. This can also feed back into programming choices, from featuring popular journalists and presenters in current affairs programming to enhancing story lines for favorite characters in drama series. Such approaches may also seek to explore the use of sentiment analysis, not only to determine the volume of mentions for specific themes or persons but also to identify the tone and context in which they are mentioned (Is a reality TV contestant controversial or popular? Is the coverage of a topic appreciated or criticized?); it should be noted in this context, however, that the effectiveness of current sentiment analysis techniques in processing the very short texts of social media posts remains disputed (Liu, 2012; Thelwall, 2014).

Context-Sensitive Approaches
to Measuring Audience Engagement

Perhaps as a reflection of the persistence of ratings thinking in social media audience measurement, existing approaches by commercial research enterprises to the analysis of social media data around television are largely based on relatively simplistic volumetric measurements. Nielsen, for example, uses the SocialGuide platform to rank shows according to what the company terms the "Unique Audience" of a show, that is, the estimated number of Twitter users who could have seen a tweet about a show (Nielsen Social, 2014). But this measurement fails to account for the different contexts in which shows air: For example, in the United States, it compares shows screening on the less subscribed USA Network to those broadcast by the mainstream national network ABC and places moderately popular FOX afternoon sporting events on an equal footing with prime time pay-per-view wrestling broadcasts. UK operator SecondSync, which has now been purchased by Twitter, Inc., similarly ranks social media activity by two volume-based metrics: total tweets and tweets per minute; in both cases, it also compares shows on different types of networks without accounting for their underlying differences (SecondSync, 2014).

Such approaches to social media audience metrics are clearly and significantly limited in their ability to measure engagement effectively. For instance, a simple ranking of shows by the total number of tweets they have received ignores the number of tweets posted per user and thus fails to differentiate between, on the one hand, broad but shallow engagement by a large number of moderately committed viewers and, on the other, deep but narrow engagement by a dedicated niche audience of fans. These generic metrics also implicitly assume that the mode of engagement with a show is the same for viewers of all formats; that is to say, they assume that audiences engage in the same way with a reality TV show as they do with a drama, for instance. But this is disproved by SecondSync's own data, which show that the peak of audience activity for drama broadcasts often occurs after the conclusion of an episode, whereas for reality TV, viewers are more likely to tweet during a show (Dekker, 2014). Although a ranking of shows by their tweets-per-minute average may allow for such genre-specific variations in audience engagement, it does not incorporate any evidence of sustained engagement with a show; a show that flatlines except for a moment of major social media controversy would rank highly by this metric, compared to a broadcast that receives solid and steady engagement throughout.

Metrics that seek to quantify sustained audience engagement, and do so with regard to the specific characteristics of that engagement, would then already be a significant improvement over currently available measurements. When seeking to understand the social media footprint of television shows, it is important that contextual factors that affect social media users' engagement with television content are accounted for. In particular, it would be desirable to normalize available measures of the volume and dynamics of content posted through social media, and thus of social media engagement with a show, by accounting for underlying systemic factors such as the geographic reach of a broadcast network, the weekday and month of a broadcast, the broadcasting genre, or the show's time slot. In this way, viewer engagement with a high-budget prime time drama on a major television network could be benchmarked more meaningfully against the social media activities around a reality TV show airing on cable television. Rather than simply comparing raw volume figures, which will always favor major channels and prime time broadcasts, comparisons could thus be based on measurements of a show's social media performance relative to the long-term average for engagement broadcasts on the same channel, in the same time slot, and/or of the same genre.

Given that this critically depends on accounting more comprehensively for the broadcast context of a given show, it is logical to consider other fields in which contextualizing statistics is significant. Noteworthy new impulses for the further development of social media engagement analytics come from the field of sports metrics, where data analysts have long faced a similar challenge to that underlying audience measurement: separating the signal from the noise (Silver, 2012). Sporting analytics has addressed this challenge by seeking to account for the fact that traditional measures of team performance (wins and losses) and players (individual statistics) can be influenced by a wide range of factors beyond the skill level and performance of a player on the field, including the skill of other players on a team's roster, the standard of the opposition, and the playing conditions of a specific match.

Despite recent developments in ice hockey, basketball, and American football (Moskowitz & Wertheim, 2012), as well as soccer (Anderson & Sally, 2013), baseball analytics remains the most developed of these fields, through the work of researchers such as James (1982), Silver (2003–2009), and Tango, Lichtman, and Dolphin (2007). The field of baseball analytics that has emerged from their efforts is called Sabermetrics (named after the Society for American Baseball Research, SABR); we therefore refer to our adaptation of these methods to the study of television audience engagement on social media as *Telemetrics*.

For the purpose of interpreting and improving contemporary audience engagement metrics, the most useful sporting analytics are those that seek to separate a player's actual performance from the contextual factors outside the player's control that may have affected it. In baseball, pitchers have historically been evaluated through a statistic called ERA, or Earned Run Average, which is calculated by dividing the number of earned runs[1] conceded by the number of innings pitched. However, this metric has been shown to be inferior to contemporary, context-based metrics. One measure of the validity of a statistic that evaluates performance is the extent to which it is predictive of future performance. However, research has shown that ERA (Swartz, 2012), as a measure of pitching ability, is not as predictive of the pitcher's future performance as those metrics that account for context. A number of competing statistics have been developed that account for particular elements of the pitcher's context, such as the quality of the fielders, the random distribution of errors, and the performance of the opposition batters whom the pitcher faced on a given day. These alternative metrics include measures such as xERA (expected ERA), FIP (Fielding Independent Pitching) and xFIP (expected Fielding Independent Pitching). The statistical measure that is most commonly used in contemporary Sabermetrics is SIERA (Skill-Interactive ERA), which measures pitching ability by taking into account only those metrics that are solely under a pitcher's control.

In measuring television audience engagement through social media, it is vital to control for the systemic boost in social media activity caused by a broadcast's time slot, network, and other factors. To do so, we can draw on a range of sporting metrics that account for such factors by weighting the standard measures accordingly. PERA, or Peripheral ERA (Baseball Prospectus Team of Experts, 2004), is one example of this: It recognizes the inherent "park factors" of each stadium where baseball is played. Essentially, this is calculated by benchmarking each playing statistic for the home and away teams in a given stadium against their overall performance away from that stadium: For example, a stadium with a home runs park factor of 112 sees 12% more home runs than the average stadium. Each pitcher's PERA can then be calculated by adjusting the counts of hits, walks, strikeouts, and home runs that underlie the standard ERA measure by the park factors of the stadium where the game was played, thus eliminating any such location-specific contextual factors.

Translating this methodology to the measurement of social media activities relating to broadcast content, we have developed a similarly context-independent metric to quantify Twitter-based audience engagement, the Weighted Tweet Index (Woodford & Prowd, 2014). Using this approach, we have been able to identify a number of the contextual factors that influence social media activity levels, including the multiplier effects resulting from the specific television network, the genre, the time of day and year, and the location of a specific episode within the seasonal cycle of a show. The Weighted Tweet Index builds on large longitudinal datasets for a wide range of U.S. television series during the 2012–2013 broadcast seasons, including Twitter activity metrics published by Nielsen SocialGuide and data collected directly from the Twitter API. Drawing on data for 9,082 individual episodes over 21 months (April 2012–January 2014), we calculated a range of contextual broadcast factors analogous to the park factors described for PERA, allowing us to understand the influence of these factors on the volume of social media audience engagement. These factors are normalized to an index value of 1; thus, a factor of 1.28 represents overall engagement 28% above average.

Unsurprisingly, the largest influence on social media engagement observed in our dataset was the broadcast channel itself: We identified a significant difference in the baseline social media activity levels for shows aired on major networks (e.g., CBS) and those shown on cable channels such as MTV. Our data contained shows on 161 U.S. television channels, with major networks such as ABC (1.15), CBS (1.09), and NBC (0.80) differing substantially from cable channels such as BBC America (0.09), Nickelodeon (0.12), and VH1 (0.47). A second key factor that influences engagement with shows on social media is the time at which an episode airs. This affects the size of television audiences more generally: Networks have defined seasons for new shows; pause shows during the winter holidays, when audiences traditionally fall; and rarely air prime shows on Fridays. Quantifying the differences between these times is key to both evaluating historic engagement values and predicting future activities; in our analysis (Table 9.1), the factors for monthly engagement varied from 0.587 (April) to 1.286 (January), and daily variation ranged between 0.24 (Friday) and 1.51 (Tuesday).

Table 9.1: Selected underlying broadcast factors affecting social media audience engagement in the United States

Month	Index Factor	Day	Index Factor	Network	Index Factor
January	1.29	Monday	1.45	ABC	1.15
February	1.10	Tuesday	1.51	CBS	1.09
March	1.20	Wednesday	1.18	ESPN	1.05
April	0.59	Thursday	1.16	FOX	0.96
May	0.62	Friday	0.24	NBC	0.80
June	0.87	Saturday	0.58	TNT	0.69
July	0.86	Sunday	0.88	MTV	0.56
August	1.03			VH1	0.48
September	0.97			ABCF	0.36
October	1.27			MTV2	0.30
November	1.10				
December	1.26				

These long-term factors are exceptionally valuable for any attempts to move beyond a simplistic ranking of shows based on their raw social media activity metrics: For the first time, they enable a benchmarking of the social media-based audience engagement with television content that is able to compare prime time and daytime broadcasts, mainstream and cable content, drama and reality TV genres without merely coming to the obvious conclusion that mainstream content generates more tweets, likes, and comments. The Weighted Tweet Index provides a valuable starting point for advancing beyond the basic metrics generated by commercial analysts such as Nielsen SocialGuide and SecondSync and constitutes a key tool for the evaluation of shows on a like-for-like basis and for predictions of how a successful cable show might fare if aired on a mainstream network. Its weighted metrics allow networks and producers to benchmark their shows against others, not just on raw numbers but by controlling for the other factors that influence audience engagement. However, it is important to note the limitations of this approach. Key among these is that such weightings can never account for the content of a specific episode. For example, in the 2013 season of *Big Brother* (United States), we saw a large spike in social media activity that was attributable to a controversy over racism; such acute events are impossible to account for through purely quantitative approaches. Necessarily, the existing weightings can also be further refined, just as the sporting analytics frameworks we have drawn from were developed over a number of years.

A particular focus of related sporting analytics has been the prediction of future performance, on both the team and the player level, for a variety of purposes. Team executives need to make decisions on roster composition, contract values, and other issues; sporting media and fan sites are tracking the performance of teams and seek to contribute insightful commentary; participants in fantasy sports and gambling markets may have significant financial investment in players' performances—these all subscribe to data sites that offer performance predictions based on cutting-edge data analytics approaches. One example of this is Baseball Prospectus's PECOTA (Player Empirical Comparison and Optimization Test Algorithm), which uses advanced Sabermetric statistics to predict players' performances several seasons into the future. These player-level statistics can then be used with the Pythagorean expectation formula, developed by Bill James (1980), to estimate the games a team *should* have won, to calculate expected wins and losses for teams over the course of a season.

By determining the contextual broadcast factors that influence social media engagement and applying them to the long-term social media engagement averages for a show once the scheduling of upcoming episodes is known, it is similarly possible to generate predictive measures of the expected social media volume for these episodes. Predictive measures can serve a number of purposes: For the viewer, they enable the selection of shows that are likely to have an active social media audience to engage with; for broadcasters, television producers, and social media strategists, they provide a benchmark to measure whether a show has been as successful on social media as it should have been; and for advertisers, they offer a tool for more targeted promotions, both through traditional commercials and directly through social media-based advertising that reaches a specific social media demographic. Although current social media audience measurement systems remain imperfect and are as yet unable to meet all of the demands of all of the various stakeholders and interested parties—producers, broadcasters, advertisers, advertising sales agents, media buyers and planners, audience research agencies, academics, and audiences themselves—they can nonetheless already illuminate new forms of audience behavior and provide insights into particular audiences' levels of engagement with screen content.

Our new approach draws on developments in sports metrics to create a method for a comparative measurement of the performance of particular television content in terms of audience engagement through the computational analysis of social media data. Our findings to date indicate that, for the moment at least, social media-derived television metrics are no cure-all for the current shortcomings of traditional television audience metrics. Ratings systems for commercial television will continue to be used for as long as the various stakeholders are able to extract value from them. New measurement systems such as Telemetrics that are based on social media analysis are unlikely to replace the ratings; rather, such systems will coexist with and complement each other as the media industries' long quest to understand their audiences continues.

Note

1. Earned runs differ from total runs in that they exclude any runs given up after a fielding error prevented the third out of an inning.

References

Anderson, C., & Sally, D. (2013). *The numbers game: Why everything you know about soccer is wrong.* London, UK: Penguin Press.

Andrejevic, M. (2007). *iSpy: Surveillance and power in the interactive era.* Lawrence: University Press of Kansas.

Andrejevic, M. (2013). *Infoglut: How too much information is changing the way we think and know.* New York, NY: Routledge.

Balnaves, M., O'Regan, T., & Goldsmith, B. (2011). *Rating the audience: The business of media.* London, UK: Bloomsbury Academic.

Baseball Prospectus Team of Experts. (2004). *Baseball prospectus.* New York, NY: Workman Publishing.

Berry, D. (2011). The computational turn: Thinking about the digital humanities. *Culture Machine, 12.* Retrieved February 25, 2015 from http://www.culturemachine.net/index.php/cm/article/view/440/470

Bourdon, J., & Méadel, C. (2014). Introduction. In J. Bourdon & C. Méadel (Eds.), *Television audiences across the world: Deconstructing the ratings machine* (pp. 1–30). Basingstoke, UK: Palgrave Macmillan.

Bruns, A., Burgess, J., & Highfield, T. (2014). A 'big data' approach to mapping the Australian Twittersphere. In K. Bode & P. Arthur (Eds.), *Advancing digital humanities* (pp. 113–129). Basingstoke, UK: Palgrave Macmillan.

Coleman, S. (2003). A tale of two houses: The House of Commons, the Big Brother house and the people at home. *Parliamentary Affairs, 56*(4), 733–758.

Council for Research Excellence. (2010). *The state of set-top box viewing data as of December 2009.* Retrieved April 26, 2015 from http://www.researchexcellence.com/files/pdf/2015-02/id143_stbfinalreport_3_5_10.pdf

Dekker, K. (2014). *2014—The top tweeted shows so far.* SecondSync Tumblr. Retrieved January 8, 2015 from http://secondsync.tumblr.com/post/80172270097/2014-the-top-tweeted-shows-so-far

Fiske, J. (1992). Audiencing: A cultural studies approach to watching television. *Poetics, 21*(4), 345–359.

James, B. (1980). *The Bill James abstract.* Self-published book.

James, B. (1982). *The Bill James baseball abstract.* New York, NY: Ballantine Books.

Liu, B. (2012). *Sentiment analysis and opinion mining.* San Rafael, CA: Morgan & Claypool.

Moskowitz, T., & Wertheim, L. J. (2012). *Scorecasting: The hidden influences behind how sports are played and games are won.* New York, NY: Three Rivers Press.

Napoli, P. (2003). *Audience economics: Media institutions and the audience marketplace.* New York, NY: Columbia University Press.

Nielsen Social. (2014). *Nielsen SocialGuide Intelligence.* Retrieved January 8, 2015 from http://www.nielsensocial.com/product/social-guide-intelligence/

Pesce, M. (2006). *You-biquity.* Keynote address from the Webdirections South Conference, Sydney, Australia.

Ryall, J. (2014, 16 July). How Netflix is quietly thriving in Australia. *Sydney Morning Herald.* Retrieved January 8, 2015 from http://www.smh.com.au/entertainment/tv-and-radio/how-netflix-is-quietly-thriving-in-australia-20140716-ztirm.html

Savage, P., & Sévigny, A. (2014). Canada's audience massage: Audience research and TV policy development, 1980–2010. In J. Bourdon & C. Méadel (Eds.), *Television audiences across the world: Deconstructing the ratings machine* (pp. 69–87). Basingstoke, UK: Palgrave Macmillan.

SecondSync. (2014). *Twitter leaderboard.* Retrieved January 8, 2015 from https://secondsync.com/leaderboard.html

Silver, N. (2003–2009). Nate Silver author archives. *Baseball Prospectus.* Retrieved January 8, 2015 from http://www.baseballprospectus.com/author/nate_silver

Silver, N. (2012). *The signal and the noise: Why so many predictions fail—But some don't.* London, UK: Penguin Press.

Silvey, R. (1951). Methods of viewer research employed by the British Broadcasting Corporation. *Public Opinion Quarterly, 15*(1), 89–94.

Swartz, M. (2012). Are pitching projections better than ERA estimators? *Fangraphs.* Retrieved January 8, 2015 from http://www.fangraphs.com/blogs/are-pitching-projections-better-than-era-estimators/

Tango, T. M., Lichtman, M. G., & Dolphin, A. E. (2007). *The book: Playing the percentages in baseball.* Lincoln, NE: Potomac Books.

Thelwall, M. (2014). Sentiment analysis and time series with Twitter. In K. Weller, A. Bruns, J. Burgess, M. Mahrt, & C. Puschmann (Eds.), *Twitter and society* (pp. 83–96). New York, NY: Peter Lang.

Woodford, D., & Prowd, K. (2014). *Everyone's watching it: The role of hype in television engagement through social media.* Paper presented at the Social Media and the Transformation of Public Space conference, Amsterdam, Netherlands.

World Federation of Advertisers (WFA). (2008, June). *Blueprint for consumer-centric holistic measurement.* Brussels, Belgium: World Federation of Advertisers. Retrieved January 8, 2015 from http://www.wfanet.org/media/pdf/Blueprint_English_June_2008.pdf

Wurtzel, A. (2009). Now. Or never. An urgent call to action for consensus on new media metrics. *Journal of Advertising Research, 49*(3), 263–265.

Staging the Subaltern Self and the Subaltern Other: Digital Labor and Digital Leisure in ICT4D

Radhika Gajjala, Dinah Tetteh, and Anca Birzescu

The question of the subaltern and representation raised by postcolonial scholars following Gramsci's (Hoare & Smith, 1999) use of the term in his *Prison Notebooks* is reexamined in the context of contemporary Web 2.0 space in attempts to understand issues of access, voice, and staging of the Other. The original questions from the subaltern studies collective were about academic and nationalist representations and involved examining strategies such as writing in reverse as a way to recover unrecorded stories (Beverley, 2004). In the Web 2.0 context, however, the question of the subaltern shifts to an examination of staging and access of economically marginalized populations. In the context of Internet-mediated globalization and financialization of the everyday, with its prevailing rhetoric of inclusivity, participation, and global access, these questions surface through philanthropy and nonprofit and corporate social responsibility groups.

As many of us continue to research issues of the digital divide and information and communication technologies for development (ICT4D), we are quick to express excitement at the mere visibility of recognizable *difference* of any kind represented in online space. We end up celebrating as unproblematic the top-down diffusion of innovation from Global North to Global South. Thus we confuse the idea of subaltern voice with the imposition of structures and technologies that adopt frameworks and design logics developed through neocolonial hierarchies that have not been adequately unpacked or examined in honest relation to subaltern everyday contexts. We are quick to point out that the subaltern can indeed speak because of digital access—we have visual evidence in the staging of the obvious. "See? They hold the gadgets," we say, "and they look happy." Thus we see a "reconfiguration of the body-as-data-body and of the political as bio-political" (Kunts-man, 2012, p. 9) that most often relies on surface (visual) staging and quantification of access points devoid of context or historicity. The digital subaltern 2.0 produced in this manner is mistaken for and conveniently used to represent the historically subalternized populations of the world.

The Internet and other ICT4D tools, including the mobile phone, thus *appear* to provide platforms for a seeming level playing field (resonating with market-economy assumptions and policy making) for the young and old, rich and poor, urban and rural, developed and developing. Yet the very tools that provide access also function to entice/lure the subaltern into providing free labor in play-like forms (digital leisure) such as commenting, liking, retweeting, and sharing news and images about specific brand names and products through the promise of economic access and empowerment. No doubt this is because Web 2.0 and mobile-based economies work through a participatory culture and prosumer paradigm, the success and maintenance of which rely on play-like forms of prosumption, leisure publics, and free labor. When the subaltern gains access, he or she then encounters these mediated cultures of interaction as given and faces the need to reorient his or her performance of self. The subaltern, therefore, is staged through online textual and visual formats, projecting an image of an empowered individual or group previously excluded or marginalized.

What is not apparent is the work the subaltern must do to appear em-powered. Setting up any presence in any online social network takes more than merely a computer and access to the Internet. Layers of literacies and skills are required before we can build ourselves into being in these spaces. As technology becomes more and more absorbed into everyday communica-tion, the individual's opportunities to contribute to a local/global culture and economy are determined by the skills known and displayed at the interface.

To produce a profile in a social media environment and engage in that context, the subaltern citizen must have Internet connectivity and basic hardware and software access and must be fairly well versed in the languages coded into the particular platform. Further layers of skills and translation include cultural knowledge of how the subaltern's visual appearance might be read in a global context. How might this individual produce an appropri-ate profile for the context? What etiquettes prevail in the social hierarchies he or she will enter? Even assuming that the subaltern citizen directly has the potential to participate, he or she must put in layers of translation work and self-skilling time to participate within this global so-called level playing field. In actuality, however, in the contexts we write of in this chapter—Kiva.org on the one hand and M-PESA marketing materials on the other—

the subaltern presence is, most often, produced through the assistance of development workers. The digital subaltern 2.0 is thus a product of layers of mediation.

However, in common discourse, connectivity to computers is itself seen as a solution to the lack of access to structures of economic and social empowerment. Inclusion is framed through approaches such as One Laptop per Child (OLPC). Even critical scholars are susceptible to the lure of thinking that just because we can easily avail ourselves of the advantages of Internet and mobile interactivity, this must somehow be wonderful for everyone else in the world. In what follows, we problematize the issue of subaltern access to the global through digital platforms by examining some examples chosen from online microfinance (Kiva.org) and mobile money (mpesa.in). These two sites have been chosen because of the global acclaim they have received.

In past and ongoing collaborations, we have analyzed online textual and visual materials to examine how the racialized subaltern is staged in digital financial platforms such as those that market mobile money projects (e.g., M-PESA in Kenya) and those that offer microlending and microborrowing opportunities (e.g., Kiva). We have argued that this digital subaltern 2.0 is an enactment of the so-called subaltern presence in global digital space and forms a portal for the emergence of a particular decontextualized (staged with selective context and background narration and images), individualized, global labor force (Gajjala, Tetteh, & Yartey, 2014). These virtual enactments, however, are *real* in their material impact and consequence: They work to reconfigure access as access to a homogenized global financial and market space. Central to our observations of these enactments is Gajjala's (2013) argument:

> What appears to be connectedness may be mere re-presentation, what appears to be exoticization may be necessary marketing for survival. What appears as individual authentic voices are thus voicings that are produced performatively through an interaction of invisible interface design and political, economic and discursive hierarchies that have coded the subaltern as data. (p. 16)

We begin our analysis with a discussion of the digital subaltern 2.0 concept, including how the subaltern is staged in the global digital space using ICT4D tools. Next, we discuss popular discourses about how development goods and services are marketed to the rural poor and how the rural poor are seen both as beneficiaries of development benevolence and as a viable market for corporations. We then provide specific examples from Kiva and Safaricom/M-PESA to explain how the subaltern is staged as Other, as Other transitioning into Self, and as Self.

Digital Subaltern 2.0

Digital subaltern 2.0 (Gajjala, 2013) is the term used to describe inclusion of the subaltern of the Global South in local and global discourses using Web 2.0 tools. Through Web 2.0 spaces and ICT4D tools, the subaltern is shown as being provided the tools to lift himself or herself out of poverty by the bootstraps and become a global citizen. These byte-sized representations of the so-called authentic subaltern in online philanthropy websites and digital giving platforms are facilitated through Web 2.0 platforms (philanthropy 2.0) and require the rearticulation and shifting of geographically and socioculturally located identities in relation to global market-economy-driven ideologies of inclusion. The larger structure does not shift toward the Other; the Other must transform and uproot from his or her familiar ways of being to be included. Therefore, the representation and staging of the Other is done such that global users and consumers can recognize the product associated with the subaltern through globally recognized stereotypes.

We refer to such representations and self-stagings as digital subaltern 2.0 avatars. These avatars begin to represent subaltern citizen (Pandey, 2010) participation and subaltern agency. Even though Internet connectivity to the global seems to allow a shift in the epistemic framing of the subaltern as exotic Other through a repositioning of subaltern as agent—as Self rather than Other—the digital subaltern 2.0 still reinstates colonial hierarchies through the use of textual and visual cues. As detailed later, Web 2.0 spaces that appear to break down barriers between the poor and the not-poor are still reinscribing colonial hierarchies.

Development Discourse and Marketing Paradigm

The Internet is both a culture in itself and a cultural artifact produced by people with contextually situated goals and interests. One particular aspect of Internet architecture is the ability to informationalize through data mining and coding. Narrative structures are also appropriated into this architecture of informationalization. Peter Chow-White (2008) argued that this results in an informationalization of race:

> As communication technologies play an increasingly centralized role in the everyday practices and organization of a range of social institutions and industries, there are a number of sites where we can see the informationalization of race at work, such as law enforcement, biomedical research, insurance, and marketing. While each would have their own set of technologies for information storage,

classification, and surveillance, they have increasingly employed a similar array of technologies to their own institutional needs and goals. (p. 1174)

Contemporary development discourse has moved into an informationalization framework and mobilizes the subaltern as product about which data (information) must be provided. This marketing is increasingly advocated as a value-building move to enhance demand for a product. Not only is access to information emphasized as a path toward development and implicitly as a cure for poverty, identities are narrated—produced visually and textually through a process of informationalization whereby they become points of reference as evidence of reality in and of themselves.

The informationalization of marginalized bodies is made possible by the free labor provided by residents of developing world communities through online leisure-oriented and community-building activities. This participatory and collaborative way of working to provide access contributes even more to the illusion of a digital level playing field. Such seeming equalization of access allows a shift in development discourse that calls for development projects to view and include what is called the "Bottom of the Pyramid" (BoP) poor populations as consumers and active agents in the development agendas for the Global South.

Development discourse can no longer adopt the traditional top-down stance where the rich empower or give to the poor. This is where philanthropy 2.0 feeds into what we might call "development 2.0," or an attempt to use participatory technologies for inclusion. In this paradigm, development workers must speak of inclusion and self-empowerment and the dignity of the poor. They now claim to facilitate independence and entrepreneurialism through strategies of collaboration and participation. Rather than speak of giving voice, they now address the self-empowerment and entrepreneurial spirit of potential slumdog millionaires. As noted by Kuriyan, Nafus, and Mainwaring (2012), ICT4D tools such as mobile phones both serve the profit interests of corporations and promise upward mobility for the poor. Corporations have realized the potential of this tool, as evidenced in the uptake of programs and promotions delivered via the mobile phone in developing countries.

In the context of digital global environments, digital play and leisure-seeming activities in social media environments constitute practices of participation and inclusion. Marketing paradigms now routinely use inclusion and crowdsourcing to mobilize and exploit existing user groups and consumers who engage in social activities around particular brand products and media texts. ICT4D frameworks are also turning to youth-based, digital-native-informed, social spaces and practices in their development campaigns and in marketing technologies to the BoP populations. Technologies are marketed as materially empowering—offering opportunities for individual entrepreneurial activity—while also promising the subaltern opportunities for inclusion in global digital leisure activities such as hanging out on Facebook or exchanging Instagram selfies, regardless of whether they actually own the hardware and software needed for access to these social spaces. Such campaigns simultaneously offer the possibility of entertainment and empowerment to the user in a seemingly nonlaborious way.

Stages of Subalternity

Our analysis in this section continues an investigation of the dataization of subaltern bodies through a process of informationalization via narrative and visual data gathered by various human actors as it reinscribes colonial hierarchies in visual Internet space. We do this through close visual and textual analysis. In what follows, we show three particular formats of representation: the subaltern at work as a hardworking, productive Other; the subaltern as Other transitioning into (framed as) subaltern as Self; and the subaltern as Self at play, engaging in globally recognizable youth-culture-based leisure activities.

Our analysis is based on examples from two platforms for the empowerment of the developing world citizen. Kiva is a nonprofit organization providing microloans by connecting philanthropists with borrowers who aspire to be entrepreneurs. M-PESA is a mobile money system that is successful in Kenya but marketed through the Internet (and by the service provider Safaricom) to Kenyans of all economic backgrounds and showcased to the rest of the world as a successful ICT4D. The customers and users of these two platforms may overlap but are not identical. However, the audience for both the marketing and promotions for M-PESA and for the Kiva borrower profiles consists of Westernized philanthropists.

Kiva displays a philanthropic discourse. Its audience is the potential lender from the Global North and its most often used strategy positions the subaltern as Other. M-PESA marketing, on the other hand, is based on the need to spread the use of mobile phone services and digital banking to urban and rural youth of the developing world. Because the goal of the persuasion is to invite the subaltern into participatory communities through the use of mobile gadgets, this organization often employs a strategy that positions the subaltern as the Self who is in pursuit of the upwardly mobile leisure consumption practices. However, on both platforms, there are instances when the subaltern is depicted as neither Other nor Self but somewhere in between, hence subaltern as Other transitioning into Self.

Subaltern 2.0 as Other

The idea of the subaltern as Other manifests where the poor of the developing world are staged as different from the target audiences of Kiva and Safaricom (through M-PESA) in terms of financial self-sufficiency, technological know-how, and social status/class. For instance, in some M-PESA promotional materials, we see a presentation of the subaltern as Other who needs to be helped out of poverty through financial assistance from relatives. In one M-PESA ad, a young urban man sends money to an older female relative living in the country. Despite being older, the woman is not daunted by technology; she is both included and empowered by technology. Despite their age differences, these two can connect via a shared technological platform—mobile money. The mobile technology is presented as empowering for the woman, because this activity would not have been possible prior to the advent of mobile technology and because through this technology use, she can become an individualized Self. But in order for the young man to continue supporting the older relative, he needs to be made aware of his position of status, which is better than that of the recipient, a status highlighted in this ad.

Other illustrations of the subaltern as Other can be seen in the presentations of borrowers on the Kiva platform. Previous work (Gajjala & Birzescu, 2010) attended to the processes of emergence of voice from marginalized groups on Kiva.org in view of its implications for existing and emerging structures of power against the backdrop of a global economy. Instead of "voice," we found the term "voicings" more suggestive of the processes taking place on the Kiva interface, because voice represents a construct based in contingent speech acts shaped through existing power hierarchies. These power structures discipline, more or less subtly, the marginalized speaker once he or she emerges as a speaking agent by fixing voice within frameworks recognizable within hegemonic mainstream discursive logics.

In the philanthropic context of Kiva, the voices of the subaltern Other are mainstreamed, removing any subversive or resistive potential so that the Western, capitalist imagination perceives their difference as nonthreatening and feels in control. The resulting relationship of subordination, whereby the subaltern Third World Other becomes exposed/known/knowable to the Western subject, is tributary to the history of colonialism and imperialism.

One of the many profiles on Kiva.org that instantiate the subaltern as Other is that of Razia from Mandi Bahauddin, Pakistan. The profile narrative presents the 28-year-old Razia and her fruit-seller husband. The husband is presented as the breadwinner, and the text emphasizes the woman's nurturing role: "Razia is a mother of six children. She is responsible for their food, health and education, and is trying hard to fulfill their basic necessities through her husband's business" (Razia, n.d.). Although she doesn't play the key role in the business, Razia seeks funding so they can purchase more produce and her husband can "meet market demands consistently" (Razia, n.d.). Her endeavor can be explained by the fact that women make up the main target of microlending practices.

Wearing a scarf, Razia is photographed with her husband behind their fruit stall against a humble backdrop of old painted brick walls and a gloomy entrance into a dark room. Their poses—Razia with her arms lowered and hands crossed and her husband with his arms hanging straight alongside his body—suggest a disciplined, nonthreatening attitude.

Certainly, in any other context, the picture of Razia and her husband might be decoded differently. However, the context adds a connotative layer. The photo and the accompanying text on Razia's Kiva.org profile are meant, like all borrowers' profiles in this global digitally mediated space, to be perused and scrutinized by any Internet viewers who land on the Kiva website and who may subsequently become lenders. Read from this perspective, the photo doesn't simply present a couple displaying fresh produce. For the gaze of the Western global lender, Razia and her husband become signifiers sharing the same implied meanings underpinning the fruits—displayed for purchase and consumption.

Borrowers' profile photos are directed to a significant extent by the Kiva fellows on the field who take the pictures (Schwittay, 2014). In this sense, we might further argue that the encapsulation of the couple in the discursive economy of the profile photograph echoes the analogy Rey Chow (2002) drew between "the predicament in which those who are labeled ethnic find themselves in white capitalist societies" (p. 96) and art historian John Berger's (1980) account of the public zoo as a seemingly innocent imperial-ist institution. In the public zoo, the intended encounter between the Western man and the caged rare animals is impossible because both the animals and their spectators "presume on [the animals'] close confinement" (p. 22). It is an environment where "the visibility through the glass, the spaces between the bars, or the empty air above the moat, are not what they seem" and where "visibility, space, air, have been reduced to tokens" so that the animals "become utterly dependent upon their keepers" (pp. 22–23). Chow (2002) argued that "However well intentioned a newly arrived onlooker may be and however much concentration she may wish to give to those inside the cage, something will, under these circumstances, always seem out of focus" (pp. 96–97).

Indeed, the majority of the profile images on Kiva perpetuate an ethnicized representation of the subaltern Other. However, the strategy of ethnicizing or Westernizing particular profiles relates to the type of entrepreneurial venture. Ethnicized representations often have more market value and attract more lenders.

Framed as Self

Since Gajjala and her research team (see Gajjala et al., 2009) began writing about Kiva a few years ago, there is clear evidence of a shift in how the subaltern is staged. Even given this shift, we see many profiles staging the subaltern as Other in a manner analogous to the ubiquitous before and after pictures in weight-loss advertisements proclaiming the individuals' transformations. The transition from the subaltern as Other to subaltern as Self is shown as a transformation from local to global identity staging, where the global identity is the ideal Westernized, self-empowered, individual entrepreneur.

Thus, there are variations in how the subaltern is presented in these online spaces; we also see a shift in the presentations. The subaltern is no longer presented in abject poverty and in tattered clothes, not only because that is not the actual situation in most cases, but also because such presentations are not appealing and can alienate the target audience. Lenders also want evidence that the subaltern is productive and making progress toward lifting himself or herself out of poverty, hence a shift in presenting the subaltern as Other transitioning into Self (framed as Self).

The workings of discourse and practice on Kiva produce the subaltern as needing to be empowered through digital technologies and mobile gadgetry. The staging of marginalized identities at the Kiva interface entails gradual nuancing, shifting, and reconfiguring of subaltern images to align them with mainstream values and current cultures of global digital work and play. This is done through a clear invoking of the poor as consumer—but simultaneously the poor need help in the form of microloans.

The video titled *A Piglet Named Kiva* (Kiva, 2013), for instance, juxtaposes the rural developing world with technology use to market mobile/digital technology for the subaltern's self-empowerment. The video stages the subaltern simultaneously as the Other who needs the philanthropist to empower her and as the self-empowered consumer who has, thanks to technology, emerged as an entrepreneur. The video's subject is Alice, a young farmer and community knowledge worker from Uganda. She tells us that with the knowledge she acquired through her training, she has helped farmers treat diseases attacking their farms. The camera cuts to a training manager—also from Uganda but clearly an authority figure—who tells us about the community knowledge worker program and how it uplifts the farming communities via technology. We see local farmers using smartphones and apps to upgrade their backward-seeming farming practices before coming back to Alice, who tells us how the use of the smartphone has allowed her to become a key figure in her community. We are shown the technology's convenient features, some of the apps, and so on. The video ends with Alice petting a piglet, which she bought with the money she saved by using the mobile phone. This pig, she says as she beams at us, is named Kiva.

These and other videos and campaigns present a progression from a subjected Other whose production and display are informed by a condescending Western development gaze to a Self/agent-appearing Other that has seemingly acquired a certain degree of agency and independence. Such a subaltern citizen will indeed emerge as an ideal entrepreneur. The progression of someone such as Alice from seemingly disempowered to seemingly empowered provides evidence of the empowerment potential of Kiva loans. Borrowers are presented simultaneously as in need of help and as entrepreneurial and hardworking. They are able to use financial assistance to transform themselves from poor into working-class global citizens. The majority of borrowers pictured on the Kiva website are portrayed with evidence of their work as farmers, herdsmen, traders, and so forth; they are defined and differentiated by their work.

These presentations of the subaltern imply *progress* toward empowerment, yet the subaltern cannot quite be considered Self because he or she still needs assistance. For example, the Kiva profile of Sukartini, an Indonesian artisan who specializes in handcrafted jewelry, describes an entrepreneur lifting herself up by the bootstraps:

> My mother instilled in me the value of a "Balinese woman," including how to prepare temple offerings. At the age of 16, I studied at SMIK (Sekolah Menengah Industri Kerajinan) in Gianyar. In this school, I learned batik, ceramics and other home industries for three years. Back from school, I took an apprenticeship with a silversmith in a village where they specialized in jewelry. Today I work together with local silversmiths to promote Bali's beauty in our own way. We make different

kinds of jewelry, mostly with sterling silver, in both traditional and modern styles. (Sukartini, n.d.)

This trend of framing or staging the subaltern as Other who is upwardly mobile and in the process of transforming into the subaltern as Self is necessary to the success of Kiva. The images invoke a Westernized, individual Self. If we didn't read Sukartini's request for funding to buy silver and gemstones for her business, we might think she lived in the United States. However, Sukartini's profile reveals the contradictions inherent in the staging of the transitioning subaltern. Her Westernized photograph represents someone who looks similar to some of the lenders from mostly developed world contexts, but the description positions her as a struggling developing world citizen. Her photo makes her look independent, but the description makes her reliant on financial help. These contradictions and juxtapositions reflect a well-thought-out marketing process. The use of the first-person narrative (replacing the subaltern-as-Other third-person narrative) only seems to bypass the representational filters at work on the Kiva.org interface and simulate a self-managed online profile. For the subaltern to overcome the perception of her premodern or not-modern location and to allow herself to be perceived as an entrepreneur in whom a lender can safely invest, the transition from third person to first person is needed. Sukartini describes herself as an artist who is modern and entrepreneurial yet rooted in her heritage. Even as she is able to manage the money we loan her and deliver results as would any modern entrepreneur, she is also not just any Western woman. Rest assured the lender is still lending to an exotic Other. The affective investment (Schwittay, 2014) of the lender from the Western, modern world in this process is garnered through the promise of giving to the exotic Other. Sukartini's profile delivers on this promise by narrating an appropriately subaltern-as-Other story even as her use of the first person shows that she is transitioning into an entrepreneurial—if still subaltern— self. Both image and text allow for Kiva's Westernized audiences to believe they are experiencing a direct connection with an increasingly *individualized* subaltern. The active voice of self-representation is associated with the assertive, empowered entrepreneur who can take charge of her finances and business.

Another example of the subaltern Other transitioning into Self is an image posted on Safaricom's Pinterest board. The image shows a well-known M-PESA user (a goatherd) running alongside Western marathon runners and a baby rhino. There is, of course, humor in this juxtaposition, but this marketing also presents a juxtaposition of the subaltern as consumer (as Self) and of the subaltern as premodern and exotic (as Other), who is transforming into the modern individual through the use of M-PESA.

Subaltern 2.0 as Self

We also noticed the staging of the subaltern as Self in M-PESA promotional materials and related discourses. In these, the subaltern is portrayed as free, independent, and empowered, and thus conferred with the status of Self. For instance, the promotional video *Relax You've Got M-PESA* (Safaricom, 2012) shows a man in Western business garb seated in his car transferring money after bank hours and a woman working on a farm who uses the technology to pay her children's school fees. These clearly show how the subaltern Other can become the subaltern Self who is becoming a global individual entrepreneur through ICT4D intervention.

One of the key marketing strategies used by Safaricom to sell M-PESA is to present digital money tools as enabling worry-free, convenient, financial inclusion and access for those at the BoP—the unbanked population in Kenya. M-PESA gives people who have basic mobile phones but not bank accounts the opportunity to access financial services (Buku & Meredith, 2014). M-PESA has an online presence but operates mainly offline and through the mobile phone. Unlike Kiva.org, which is a microfinance portal and where the argument can be made about the Western world empowering the poor in the developing world, M-PESA stages self-empowerment of developing world populations.

M-PESA promotional images and videos are aspirational—they target the subaltern as much as they represent the subaltern to Westernized audiences. Images portraying the convenience of shopping online (associated with leisure activity) are presented to show how M-PESA users can become upwardly mobile. For instance, in a promotion announcing a lottery for a Toyota RAV4, the caption is a cheery "Lip A Na M-PESA at Uchumi and win a RAV 4," with five people pushing their shopping carts behind smaller images of cars. They look happy and excited—as if they have won. This image projects the ideal upwardly mobile consumer and reinforces the idea that by engaging in such practices as shopping in a Westernized grocery store, one may win further upward mobility and get closer to a globalized American dream by owning a Toyota RAV4. A social feature of upward mobility is the availability of leisure time; the ads that show how the use of mobile money allows for more leisure equate leisure with empowerment.

M-PESA marketing associates the product with convenience/leisure, progress, Westernization, empowerment, and upward mobility. Leisure is reconceptualized as time in which people can control what they do, when, how, and with whom. Considering that Kenya is located in a region of the world where online shopping is uncommon, the opportunity to shop online is empowering; the staging of subaltern as Self in this case targets the subaltern consumer and his or her aspirations. Previously, most Kenyans would not have had the option of shopping digitally; Kenyans now have a choice, thanks to M-PESA. Thus, M-PESA is simultaneously offering the leisurely act of shopping online and seeking to produce empowered consumers. Shopping online (or paying for shopping online) supposedly produces self-reliance, hence our argument that the ads imply that leisure leads to empowerment and thus a construction of the subaltern as Self.

In M-PESA marketing, the staging of the subaltern as Self is done in such a way that no distinction is made between the poor and rich or urban and rural; anyone can become upwardly mobile by working hard. This sense of a level playing field can also be seen in the online platforms where M-PESA is promoted. Social media platforms such as Instagram, Facebook, and Twitter do not differentiate the social classes of M-PESA customers; all one needs is a mobile phone with an Internet connection and one can seamlessly merge with other users across socioeconomic backgrounds. The mobile phone and other technologies bridge the gap between rich and poor. On Twitter, for instance, through activities such as tweeting and retweeting, M-PESA customers promote the product to their friends and followers. For example, an M-PESA tweet reads, "You can find the #RelaxUkoNaMPESA deck chairs at roundabouts, utility points, banks among other public areas." Customers replied to the tweet with comments such as "Been wondering what those are lol," "can i sit on them 2 relax?" "looks kinda comfy :) can I take one home?" and "so wat r we supposd to do sit on those deck chairs na hamna adabu" (Safaricom Limited, 2012). By using their cell phones to participate on social media, an activity that seems leisurely on the surface, people use their time and participation to help promote the company's products. These leisurely discussions about M-PESA can produce feelings of belonging and inclusivity among M-PESA users, regardless of their physical location and social status, hence the notion of an empowered people.

The Transition From Subaltern Other to Westernized Individual Self as Neoliberalization

Our categories of subaltern as Other, subaltern as Other transitioning into Self, and subaltern as Self are transitional—the technological interventions that the subaltern accesses will ultimately allow him or her to become an ideal neoliberal, Westernized individual Self. He or she will be transformed and will no longer be the subaltern, outside of the global economy. Our analysis and the labels we adopt reflect a continuum where the staging tells the story of progression from Other to Self, moving toward the ideal of Westernized individual agent.

This is in sync with the idea of the BoP as consumer. In such a BoP-as-consumer paradigm, microfinance and the uplifting of the unbankable poor through mobile money can no longer be framed purely in the language of charity and economic progress through hard work. We have seen this paradigm shift even in the way the low-income borrower from the Global South is portrayed/staged in the context of online microfinance. Productivity, altruism, sharing, and pride must work together to insert strains of upbeat playfulness in how the messages are conveyed. The promise of leisure and self-empowerment that makes possible entrepreneurial self-definition and control of time to work and play is implied in this framework.

In the case of Kiva, the staging of the subaltern is supposed to be non-threatening to Western lenders who must continue making financial contributions. After all, it is easier to keep giving when we know we are better off than those we support, when we know they will not rub shoulders with us anytime soon. Yet, even though this distinction between the Self and Other is necessary for philanthropy 2.0 and ICT4D frameworks, the transition from subaltern as Other to a stronger focus on the subaltern as Self can be viewed as a transition into a neoliberal economic context where individualization and privatization are privileged. Framing the subaltern as Self allows the propping up of the ideal poor person who deserves access to the global by virtue of having entrepreneurial drive and individual ambition and looking Westward culturally. This overall continuum of representation that is based in the staging of the subaltern via Web 2.0 platforms—from Other through Self—seeks to convey an image of a level playing field for the rich and poor, urban and rural, developed and developing.

Note

The coauthors want to thank Rebecca Lind for her tireless comments and feedback to help this chapter become what it is now.

References

Berger, J. (1980). *About looking*. New York, NY: Pantheon.

Beverley, J. (2004). Writing in reverse: On the project of the Latin American subaltern studies group. In A. del Sarto, A. Rios, & A. Trigo (Eds.), *The Latin American cultural studies reader* (pp. 623–641). Durham, NC: Duke University Press.

Buku, M., & Meredith, M. (2014). Safaricom and M-PESA in Kenya: Financial inclusion and financial integrity. *Washington Journal of Law, Technology & Arts, 8*(3), 375–400.

Chow, R. (2002). *The Protestant ethic and the spirit of capitalism*. New York, NY: Columbia University Press.

Chow-White, P. A. (2008). The informationalization of race: Communication technologies and the human genome in the digital age. *International Journal of Communication, 2*, 1168–1194. Retrieved April 26, 2015 from http://www.ijoc.org/ojs/index.php/ijoc/article/viewFile/221/243

Gajjala, R. (2013). *Cyberculture and the subaltern: Weavings of the virtual and the real*. Lanham, MD: Lexington.

Gajjala, R., & Birzescu, A. (2010). Voicing and placement in online networks. In M. Levina and G. Kien (Eds.), *Post-global network and everyday life* (pp. 73–91). New York, NY: Peter Lang.

Gajjala, V., Gajjala, R., Birzescu, A., & Anarbaeva, S. (2009, June). *From microfinance to online socio-business-networking: Microfinance in the age of social networking*. Paper presented at the First European Conference on Microfinance, Brussels, Belgium.

Gajjala, R., Tetteh, D., & Yartey, F., (2014). Digital subaltern 2.0: Communicating with, financing and producing the Other through social media. In R. L. Schwartz-DuPre, (Ed.) *Communicating colonialism: Readings on postcolonial theory(s) and communication* (pp. 246–266). New York, NY: Peter Lang .

Hoare, G., & Smith, G. N. (1999). *Selections from the prison notebooks of Antonio Gramsci*. London, UK: The Electronic Book Company.

Kiva. (2013, September 17). A piglet named Kiva: What Grameen Foundation AppLab is making possible in Uganda. Retrieved February 1, 2015 from http://pages.kiva.org/node/12194

Kuntsman, A. (2012). Introduction: Affective fabrics of digital cultures. In A. Karatzogianni & A. Kuntsman (Eds.), *Digital cultures and the politics of emotion: Feelings, affect and technological change* (pp. 1–20). Hampshire, UK: Palgrave Macmillan.

Kuriyan, R., Nafus, D., & Mainwaring, S. (2012). Consumption, technology, and development: The "poor" as "consumer." *Information Technologies and International Development, 8*(1), 1–12.

Pandey, G. (2010). *Subaltern citizens and their histories: Investigation from India and the USA.* New York, NY: Routledge.

Razia. (n.d.). Retrieved February 25 2015 from http://www.kiva.org/lend/843436

Safaricom. (2012). *Relax, you've got M-PESA.* Retrieved June 2014 from http://www.safaricom.co.ke/personal/M-PESA/M-PESA-services-tariffs/relax-you-have-got-M-PESA

Safaricom Limited. (2012, September 24). You can find the #RelaxUkoNaMPESA deck chairs at roundabouts, utility points, banks among other public areas [Tweet]. Retrieved February 1, 2015 from https://twitter.com/SafaricomLtd/status/250159630163001344

Schwittay, A. (2014). New media and international development: Representation and affect in *microfinance (rethinking development)* [Kindle Edition]. Taylor & Francis: Abingdon Oxon, UK. Retrieved February 1, 2015 from http://www.amazon.com/New-Media-International-Development-Representation-ebook/dp/B00O1PQJIC/ref=tmm_kin_title_0?_encoding=UTF8&sr=&qid=

Sukartini. (n.d.). Retrieved February 1, 2015 from http://www.Kiva.org/lend/755073

Race, Gender, and Virtual Inequality: Exploring the Liberatory Potential of Black Cyberfeminist Theory

Kishonna L. Gray

I'm tired of not seeing me. I'm tired of not hearing my story. Mainstream media will never get me right, so we have to take it upon ourselves to do it. Yeah, social media gives us this opportunity. (*StealMagsandNolias, personal communication, July 5, 2010*)

We can blog all we want, we can post all the pictures and memes and videos til our hearts are happy. But what are we really doing? What are we changing? Nothing. It's just more internet mess. (*TastyDiamond21, personal communication, July 5, 2010*)

Black women have varied responses when employing Internet technologies for empowerment. New communication technologies have expanded the opportunities and potential for marginalized communities to mobilize in this context counter to the dominant, mainstream media. This growth reflects the mobilization of marginalized communities in virtual and real spaces, reflecting a systematic change in who controls the narrative. No longer are mainstream media the only disseminators of messages or producers of content. Everyday people have employed websites, blogs, and social media to voice their issues, concerns, and lives. Women, in particular, are employing social media to highlight issues that are often ignored in dominant discourse (Shirky, 2011). However, access itself neither ensures power nor guarantees a shift in the dominant ideology. Many women recognize the potential of social media to improve their virtual and physical outcomes, but they also recognize the limits to which technologies can sustain a narrative counter to the current hegemonic structure. As *TastyDiamond21* suggested, regardless of how much content women create, the Internet will never have the power to dismantle society's dominant structures.

Just as with feminism and feminists, there are multiple variations of cyberfeminism (Brah & Phoenix, 2013; Crowley-Long, 1998; Mouffe, 1992; Narayan, 2013). In this chapter, I argue that Black cyberfeminism may address the critique that traditional virtual feminist frameworks do not effectively grasp the reality of all women and may help theorize the digital and intersecting lives of women (Gray, 2013). Operating under the oppressive structures of masculinity and Whiteness that have manifested into digital spaces, women persevere and resist such hegemonic realities (Gray, 2012a). Yet the conceptual frameworks intended to capture the virtual lives of women cannot deconstruct the structural inequalities of these spaces. Cyberfeminism, technofeminism, and other virtual feminisms may address women in Internet technologies, but they fail to capture race and other identifiers that must also be at the forefront of analysis.

Black cyberfeminism, as an extension of virtual feminisms and Black feminist thought, incorporates the tenets of interconnected identities, interconnected social forces, and distinct circumstances to better theorize women operating within Internet technologies and to capture the uniqueness of marginalized women.

Examining the Possibilities
and Limitations of Cyberfeminism

Cyberfeminism is useful in contextualizing the virtual nature of women's lives. Broadly, cyberfeminism is a notion that the Internet has liberating qualities that can free us from the confines of our gendered bodies (Bromseth & Sundén, 2011). The premise, however, has been criticized as both utopic and irrelevant to women's circumstances in new technologies. We cannot just forego our bodies in virtual spaces, because much of our real-world selves are emitted into these spaces. The discussion must move beyond the confines of the digital and be reexamined for its potential to mobilize women in both digital and physical spaces. The virtual and physical selves are inseparable. We must critically engage with the recursive relationship between our physical environments and our virtual selves, and we must use the framework to improve women's lives.

Women of color have long recognized that self-determination is a critical component to moving beyond the parameters of hegemonic ideology (Collins, 2000). Black feminist thought in particular argues for self-definition, a reclaiming of identity, and empowerment for all women and other marginalized groups. In this essay, I build on cyberfeminism and Black feminist thought to articulate the utility of a Black cyberfeminist framework in examining the issues that continue to impede the progression of marginalized women in media, technology, virtuality, and physical spaces.

The Internet has been touted for its liberatory promise (Magnet, 2007), but the potential for such transformation could be thwarted by attacks on women in technology. For instance, Zoe Quinn, Brianna Wu, Anita Sarkeesian, and others have been systematically targeted for being social justice warriors (Kain, 2014): #GamerGate, which began as an online movement concerned with ethics in game journalism, morphed into an attack on women and feminists. The continued sexism permeating gaming culture is part of a larger culture in technology that devalues women as full participants. This type of structural inequality is not adequately addressed by cyberfeminism. However, by incorporating a critical feminist stance, such systemic problems can be articulated while moving toward meaningful ends for women in these spaces.

How likely is it that Internet technologies can reach their liberatory potential? Many women remain on the periphery of Internet technology. Internet technologies and virtual communities are assumed to be White and masculine (Daniels, 2013; Gray, 2012a; Kress, 2009). These unequal power relations are accepted as legitimate and are embedded in the cultural practices of digital technology. But many women have resisted this perpetual state of second-class citizenship. Black feminists in particular have outlined a template for countering the hegemonic narrative often operating in technology. By blending cyberfeminism and Black feminist thought, I provide a frame to begin the discussion of allowing women to exist on their own terms and to craft their own narratives. This framework is not new, but it is distinct, given its purpose and intent. This approach details women's experiences and also provides meaningful solutions to combat inequitable power structures.

Black Women, Identity, Media, and Control

Media portrayals offer singular visions of women's lives, their behaviors, and their roles. Women are consistently underrepresented and misrepresented across various media (Glascock, 2001; Signorielli, 1997). Feminists are particularly concerned about the representations of women and femininity that promulgate unrealistic standards of physical appearance (Ward & Harrison, 2005); girls and women evaluate themselves based on these idealized representations (Field et al., 1999; Groesz, Levine, & Mumen, 2002; Levine, Smolak, & Hayden, 1994).

There are additional concerns for women of color. Television represents women of color as hypersexual, promiscuous, and immoral (hooks, 1992; Patton, 2001). Many media outlets rely on updated versions of minstrel-era stereotypes, such as the hot-tempered and loud-mouthed Sapphire, the domestic servant or Mammy, and the promiscuous Jezebel (Emerson, 2002; Stephens & Phillips, 2002).

These images are in constant clash with women's reality. Women and girls face conflicting messages about who they are, who they should be, what they can become, and how they should act (Richardson, 2007; Stephens & Few, 2007). Additionally, the racialized element inherent in mediated imagery further serves to perpetuate dominant ideology in the lives of women of color. Conflicting constructions of Black womanhood only serve to reify who is and who is not eligible for full inclusion into womanhood. Black women have long had their identities constructed by outside forces, by masculinity, and by other entities not valuing Black women's agency.

Black women and girls struggle for self-determination and self-definition against their ghettoized and distorted representations (Richardson, 2009). Hegemonic ideologies dominate the narrative of female life in the public sphere; women must work hard to resist these destructive forces. Social media have provided a means to combat these oppressive narratives and allow women the ability to define their own realities. As cyberfeminists contend, Internet technologies are an effective means to resist repressive and oppressive gender regimes and enact equality (Orgad, 2005; Plant, 1997; Podlas, 2000). However, because Internet technologies still embody hege-monic ideologies and privilege Whiteness and masculinity, the potential to resist dominating structures of oppression may be slim (Kress, 2009). As *TastyDiamond* contended, the tools afforded to women in digital spaces and in technology may allow marginalized bodies to make a temporary differ-ence, but these tools have limited ability to effect genuine change.

This concept reflects a core component of Black feminist thought. As Lorde and Clark (2007) posited, the master's tools will never dismantle the master's house. This is fundamental reality that those with consciousness recognize: The oppressed will never be given full access to spaces, websites, blogs, social media, and other Internet technologies. Although technologies were never created with the intent to destroy the hegemonic structure, they can provide temporary or partial gains in countering the establishment. And because they provide empowerment to the women who employ them, they are useful. But this compels one to ask whether the marginalized can ever truly be liberated from their oppressor. Using cyberfeminism as a starting point, it is necessary to critically examine the frameworks' limited ability to effect change.

Cyberfeminism and Technofeminism: Exploring the Tensions

From the beginning, cyberfeminism situated itself firmly in the intersections of theory, media art, and online networking (Paasonen, 2011), imploring a postmodernist belief in the interests of cyberspace, the interweb, and technology, and collapsing previously oppressive binaries such as human/machine and subject/object in a liberating action (Wajcman, 2008). The goal of a cyberfeminist perspective is to combat how women on a global scale are affected by the growing communications and technology fields and how within these spaces there can be opportunities to resist and reconstruct. Studying the various practices and engagements with this new technology takes the cyberfeminist from an examination of women in the work force (Shih, 2006) to organizing feminist political voices through online networking (Everett, 2007; Minahan & Cox, 2007). As Hawthorne and Klein (1999) framed it, "it is a philosophy which acknowledges, firstly, that there are differences in power between women and men specifically in the digital discourse; and secondly, that Cyberfeminists want to change that situation" (p. 2).

Strengthening this perspective, Brophy (2010) argued that cyberfeminism addresses the complexity of the intersection of gender and power in digital technologies, becoming a medium allowing effective resistance and equality (Daniels, 2009). Wajcman (1995) argued that the new digital technologies provide a way to destabilize the conventional gender differences established by patriarchy. Furthermore, this ever-growing space can become a place for the (re)imagination of the woman's own self, creating a projection of herself onto the intraweb (Everett, 2004). As Wajcman (2008) summed up, "young women in particular are colonizing cyberspace where, like gravity, gender inequality is suspended. As a result, our interactions are fundamentally different because they are not subject to judgments based on sex, race, voice, accent or appearance" (p. 12).

Indeed, Plant (1997) stressed that cyberfeminism is a posthuman insurrection, a new system allowing women via computers a means to resist the patriarchal world, a place where our brains matter, a location where networks, not hierarchies, operate. In this world, we are finally free from the confines of our bodies; this utopian ideal allows us to leave our bodies and to interact freely among other users. Cyberfeminism suggests that virtual spaces are locations where differences and social contexts can be erased, and our existence is defined on merit alone (Brophy, 2010; Puente, 2008).

Technofeminism: A Liberating Concept or More of the Same?

Although there have been a number of attempts to clarify the postmodern ideal of cyberfeminism, Wajcman (2004) has harnessed a number of attributes centering on gender and technology into a theory she calls "technofeminism." For Wajcman, technofeminism is a theory constructed in response to the broad scope and utopianism of original cyberfeminism, utilizing a social science approach to technology to understand the interconnectivity of gender and technology, not just in its ability to produce a reality where women can shed their worldly attributes and become anyone they desire, but as an emancipatory notion that encompasses both the positive aspects of technology and the pitfalls experienced by those women who are consumed by the modes of production. Wajcman's hope, with technofeminism, is to provide a concise guideline for an emancipatory politic in contrast to cyberfeminism's disconnection with the practical. Technofeminism differs from cyberfeminism in the sense that it is a clear and concise theory that treats technology as a sociotechnical artifact and allows us to avoid the pitfalls of an unfocused utopian idea. Technofeminism holds that technology can best be understood as both a foundation and a product of gender relations. Taken in such a context, the cultural project of technology, its purposes, and the skills required for its use all find gender to function as a means to understand how one is to negotiate such spaces (Wajcman, 1995). For technofeminism, gender is produced simultaneously with technology.

Technofeminism encompasses notions of technology in a broad sense, refusing to isolate the means of production from the technology's consumption. This notion allows feminist scholars to interpret not only the emancipatory developments some women are able to achieve but also the exploitative requirements for female workers in developing countries who are chained to the production of said technologies (Wajcman, 2006).

Still, technofeminism is not inclusive enough. Wajcman's (and others') often oversimplified, tidy narratives run counter to the feminist propensity to both/and rather than either/or. Additionally, although technofeminism emerges from third-wave feminism, it makes few claims of multiplicity, complexity, and intersectionality.

Painting these broad strokes to historicize various feminist strands is a tactic Paasonen (2002) criticized as intended to "recontextualize" the past in order "to affirm the present, implicitly suggesting the 'progress of contemporary culture' and its 'hip attitudes'" (p. 362). One charge for feminist (and other) scholarship, then, is to continue to press for ways to talk about past feminist movements and scholarship without creating a model where feminisms lose their plurality and become articulated as tales of progress, ignoring the struggles (Paasonen, 2002). Unfortunately, Wajcman sometimes falls into this trap of recontextualized, oversimplified articulation of past feminist ideas by lumping all cyberfeminists into one large group. Because of the limitations and shortcomings of both cyberfeminism and technofeminism, I urge an incorporation of a critical race feminist stance to better situate the current realities of wired women.

Black Feminist Thought in the Digital Era

Dealing with historical and contemporary oppression and marginalization, the lives of marginalized women in the digital era require an engagement with an emancipatory theoretical orientation, one that recognizes the distinctness of their shared and lived experiences. But even with the common threads woven into the patterns of women's lives, the ability to thwart the nature of dominant ideology proves daunting:

> While an oppressed group's experiences may put them in a position to see things differently, their lack of control over the apparatuses of society that sustain ideological hegemony makes the articulation of their self-defined standpoint difficult. (Collins, 2000, p. 185)

Collins (2000) outlined four perspectives unique to the standpoint of Black (and other marginalized) women: (a) self-definition and self-evaluation, (b) the interlocking nature of oppression, (c) the embrace of intellectual thought and political activism, and (d) the importance of culture. In what follows, I discuss these tenets and move toward a theoretical framework to understand the liberatory potential inherent in media and technology for women whose lived realities are reinforced through the intersecting nature of their ascribed identities.

Self-Definition and Self-Evaluation

Identity development theories address the significance of oppression in how one begins to develop identity, but the literature is limited in its examinations of multiple oppressions. The literature also fails to address the psychological impact of racism and oppression on women of color. Focusing on a specific identity or form of oppression in isolation may obscure inaccuracies in a psychological analysis of identity development in oppressed people (Smith, 1991); examining only gender or only race fails to capture the existence of women of color.

The oppressed have a unique standpoint in that they share particular social locations, such as gender, race, and/or class. Although damaging imagery of women of color permeates society, we can find evidence of contestation, resistance, and agency. These individuals share their meaningful experiences with one another, generating knowledge about the social world from their points of view. Despite this knowledge generation, oppressed populations lack the control needed to reframe and reconceptualize their realities. But particular advantages present themselves with the diffusion of information technologies: Women can create and control virtual spaces largely unregulated by the hegemonic elite. These spaces have the potential to foster the development of a group standpoint negating the impact of dominant ideology.

With these alternate spaces in place, women can begin to define their own identities and realities and influence perceptions of womanhood. They can resist the prevalence of controlling images (Collins, 2000), which reflect a system used to physically, economically, and socially control Black women. The power to manipulate images of Black women in such a way creates oppressive imagery that appears "natural, normal, [and] inevitable" (p. 5).

Interlocking Nature of Oppression

The ability for hegemonic imagery to influence perceptions exposes the ideological dimensions of women's oppression. Within this hegemonic domain of power, only ideological images characterizing marginalized women as less than human could advance and legitimate a system so fundamentally built on human degradation. This dynamic has existed since the arrival of colonists and since slavery. The images are merely recycled and remixed to further women's oppression.

Yet marginalized women consistently resist and rarely internalize these images (Collins, 2000); indeed, Black women struggle to indict the legitimacy of images such as mammies, matriarchs, welfare mothers, and so forth, as well as the integrity of those who circulate them. This process, which reveals the strong presence of a counterhegemonic consciousness, can be engaged via digital media. The relative ease with which digital spaces can be created presents oppressed groups with the ability to control and create positive content influencing our own images. For Black women, the Internet provides the potential space in which to thwart negative representations disseminated through the media (Collins, 2000).

Given the interlocking systems of oppression marginalized women experience, knowledge is especially prized for its functionality and intentionality and for its ability to help navigate and enhance one's life and community. Social networks, virtual communities, and other digital media are an extension of traditional communities, such as churches, families, and workplaces.

Embracing Intellectual Thought and Political Activism

Throughout Black liberation movements, intellectualism has been at the core of the struggle. Intellectualism simply means the knowledge that one has about who one is. It is not rooted in educational attainment. This knowledge of self propels one to the realization of liberation. Drawing parallels from the academic settings discussed by Collins (2000), Black cyberfeminist communities seek collaborations and community building among all groups working to dismantle hegemonic structures, thus highlighting the expansion—the deconstruction—of the terms intellectual and activist. Many feminists adopt an either/or approach, assuming the role of either intellectual or activist, but Black feminists urge that the space must be open for all to take equal part, existing at the intersection of intellectualism and activism.

Action and thought are not at odds but complementary. According to Collins (2000), Black feminist thought leads to Black activism; a dialogical relationship suggests that changes in thinking may be accompanied by changed actions and that altered experiences may in turn stimulate a changed consciousness. For Black women as a collective, the struggle for a self-defined Black feminism occurs through an ongoing dialogue whereby action and thought inform one another.

Empowerment and Embracing Culture

Empowerment is another important tenet of Black feminist theory. Women's power has the ability to produce transformation (Miller, 1991). Power in this sense is not about control, power over, or dominance. This power is practical; it exists in thoughts and emotions that may influence others, thus igniting interpersonal change and leading into a larger movement. Black women have historically used their power to empower others (Collins, 2000). One of the strongest movements occurring as I write this centers on #BlackLivesMatter. Having its roots in the death of young Black men at the hands of White police officers, the empowerment felt within the Black community and non-Black allies is directly linked to that interpersonal change.

In discussing empowerment, Collins (2000) noted that the center of Black women's activism "reflects a belief that teaching people to be self-reliant fosters more empowerment than teaching them how to follow" (p. 235). Personal empowerment comes in the form of self-definition and "self-knowledge as a sphere of freedom" (p. 130). The ability to be independent of the definitions set by the power structure and to produce what one wants to produce about oneself is a form of freedom.

Dei (1995) used empowerment within an indigenous knowledge frame-work, theorizing that empowerment is the self or agency having the power to voice and articulate concerns. Therefore, the possibility exists that the notion of empowerment could decentralize the hegemonic power of the dominant culture and make space to use indigenous knowledge.

Empowering one's self leads to embracing one's culture. During the Civil Rights Era of the 1960s, the call for Black power was about affirming Black humanity. White America often assumed it was the call for power and superiority over White society. Instead, it was the move to defend dignity, integrity, and institutions within Black culture. As David Chidester (1992) articulated, "it was patently not about abandoning our black communities and rejecting our black culture, but about developing the one and embracing the other" (p. 200). The same premise applies to Black cyberfeminism.

Imagining Black Cyberfeminism

Black feminism can address concerns in the virtual lives of women leading toward a critical cyberfeminist framework. Here I modify the tenets of Black feminism to reflect women in digital realms. Specifically, Black cyberfeminism concerns itself with three major themes: (a) social structural oppression of technology and virtual spaces, (b) intersecting oppressions experienced in virtual spaces, and (c) the distinctness of the virtual feminist community.

Social Structural Oppression of Technology and Virtual Spaces

Matters of institutional racism, damaging stereotypical images, sexism, and classism are routinely addressed by Black feminists (Potter, 2006). Incorporating the inherent masculine bias in technology and the default Whiteness of virtual spaces (Gray, 2012a), this theme is imperative to the creation of a Black cyberfeminist framework. Kolko, Nakamura, and Rodman (2000) argued that the Internet is far from liberatory but rather is a space that continues a "cultural map of assumed whiteness" (p. 225). Kolko (2000) pointed out that attempts to make race and ethnicity present are met with color-blind resistance. The assumed White masculine body excludes women and people of color; the mere presence of their bodies marks them as deviant in these spaces (Gray, 2012b). Deviant social behavior manifests in the materiality of the body. Blackness and any association with Blackness are punished in virtual spaces, leading to the exclusion of marginalized women.

Ignoring the diverse lives of virtual inhabitants also leads to the inability of marginalized bodies to define their own virtual realities. Marginalizing narratives perpetuated through the media reinforce limited conceptualizations of women. Black cyberfeminists urge women to regain control of hegemonic imagery, and Internet technologies allow for this. But as the limits of cyberfeminist and technofeminism illustrate, women need to ensure that they do not recreate oppressions. As I wrote this, I was reminded of an innovative form of Black cyberfeminist activism in the creation of the Twitter hashtag, #whitefeministrants. Online writer and blogger Mikki Kendall started the hashtag in response to the Whitening of feminist spaces where voices of color and otherwise marginalized women are excluded. The tweet, "My Feminism is More Important Than Your Anti-Racism: How to Properly Rank Oppression," immediately differentiates feminism by race and high-lights the privileges and oppressions yet to be addressed within the feminist community. It also highlights that traditional spaces can be co-opted, allowing marginalized women to address their grievances. In this situation, marginalized women identified the power of social media to address the racialized distinctions within feminism.

Intersecting Oppressions in Virtual Spaces

The second theme of Black cyberfeminist theory is that women must confront and work to dismantle the overarching and interlocking structure of domination in terms of race, class, gender, and other intersecting oppressions. Because individuals experience oppression in different ways, we must not create a one-size-fits-all understanding of oppression. Black cyberfeminism requires understanding the diverse ways that oppression can manifest in the materiality of the body and how this translates into virtual spaces.

Black cyberfeminism also requires a recognition of the privileges that some marginalized bodies hold before we can begin dismantling these privileges and understanding the multitude of ways that intersectionality can manifest. Such an understanding might have prevented what was referred to as the feminist Twitter war and avoided claims that Black women and other women of color lead to toxicity in virtual spaces. As Jessie Daniels (2015) explained, "the dominance of white women as architects and defenders of a framework of white feminism" that still permeates must be critically examined. So in the post proclaiming, ". . . my oppression is greater than yours," White feminists are criticized for ranking oppression, dismissing grievances of women of color and other marginalized groups. This ranking runs counter to the goals of Black cyberfeminism. Ranking oppression only leads to further marginalization of groups already on the periphery.

Black cyberfeminism, in the spirit of feminism, encourages a privileging of women's perspectives and ways of knowing, because race, gender, class status, disability, sexuality, and a host of other identifiers generate knowledge about the world. Valuing these perspectives is the only way to liberate women from the confines of hegemonic notions deeming these identities unworthy.

Black cyberfeminism also recognizes that the lived experiences of women manifest in the virtual world as well. Women do not have the luxury of opting out of any aspect of their identity. By privileging these once marginalized identities, Black cyberfeminist spaces can begin to move women toward progressive and meaningful solutions to hegemonic notions about women.

Although all women share a common struggle, examining their intersecting realities reveals the distinctness of their lived experiences. Women may share sexual oppression, but it is not clear how this can unite all women whose lives, work, life expectancy, and family life are also structured by the hierarchies of racism, ethnicity, colonialism, or nationalism.

Power differences among women are so great that even the similar struggles against men are different. Women's struggle with technology is indirectly a struggle with masculinity, patriarchy, and male privilege; marginalized women also struggle with Whiteness. Cyberfeminists' inability to incorporate the structural nature of inequality results in a limited vision of liberation. As Fesl (1984) recognized, women cannot stand together against oppression if we stand in different power relationships to one another, but as the feminist Twitter war illustrated, power differences along racial lines continue to keep feminist communities divided.

Accepting the Distinctness of Marginalized Virtual Feminisms

Black cyberfeminism also addresses the distinct nature of how women utilize virtual technologies. Women have used social media for activism and change, as well as to advance contemporary feminism. The Internet has propelled activism and empowerment in that many individuals can take action on a single issue. The tenets of Black cyberfeminism never detach the personal from the structural or the communal, which sets Black cyberfeminism apart. The key is in how marginalized women, specifically Black women, communicate and how Black women's Internet usage is a continuation of their offline selves.

Black women have used their social, cultural, historical, political, and religious reality to create their own language. Scott (2002) suggested that they even use specific words to emphasize their unique group membership. Hobbs (2004) found that the online forum of the magazine *Essence* was used to create an African American discourse space in which the cultural norms of the African American community are discussed and reproduced (p. 10). Research such as this is significant for Black cyberfeminism because it values different ways of knowing and being. Black women recognize the diverse ways of speaking, without privileging standard American English. An examination of Black women's blogs revealed their engagement with both their personal experiences and structural inequalities (Brock, Kvasny, & Hales, 2010). Conversations that once occupied beauty salons, church meetings, and kitchen tables are now present in online spaces. Black women's experiences in physical spaces influence their participation in online settings.

Black women were once touted as poster children for the digital divide. What wasn't understood was the cultural and technical savvy that Black women incorporated to use technology on their terms and for their own purposes. A technology may have been created for one purpose, but Black women will employ it to fulfill their own needs, thus displacing the hegemonic establishment.

Black women engage in a variety of cultural forms beyond traditional virtual methods of blogging or tweeting. Black women employ music (Kopano, 2002), poetry or spoken word (Johnson, 2010), and other cultural art forms in their online lives. This direct extension of the physical into the digital acknowledges the accessibility and viability of these cultural artifacts to reproduce Black feminist thought.

In addressing the accessibility of different ways of producing knowledge, Black women's engagement with art recognizes the class boundaries inherent in traditional means of cultural production. When considering gender, race and class issues must also be acknowledged: The Black feminist tradition is rooted in a belief that multiple oppressions cannot be separated, and Black feminism at its core is a strategy of resistance against the multiple oppressions of patriarchy, racism, sexism, heterosexism, and so on.

Digital social media are important in that they represent, for women of color and other marginalized groups lacking resources, a path to a space where their voices are heard. The once voiceless can be heard, and that leads to empowerment. Twitter, Facebook, and other social networking sites have allowed women to empower themselves and mobilize their communities. As Black Twitter has illustrated, people of color have co-opted traditional virtual spaces for their own means to communicate and empower their communities. So by employing the cultural tradition of sygnifyin', marginalized bodies can express themselves with others without fear of retaliation or being othered within the spaces (Florini, 2013).

Black women's use of social media also reflects their incorporation of digital technologies and their continued efforts on the ground. Twitter and Facebook have been used to organize marches, highlight continued sexism on college campuses, and draw attention to any number of issues. Maybe, in fact, because of Black cyberfeminism's simultaneous engagement in the virtual and physical communities, the master's tools will be able to dismantle the master's house.

Black Cyberfeminism:
From the Streets to the Information Highway

Black cyberfeminism, which represents the blending of multiple ideas into a cohesive analytical framework, simultaneously contributes to and widens the scope of cyberfeminism, technofeminism, and Black feminist thought. Although all three share many theoretical assumptions, values, and aims, their confluence is truly as distinct as the women who exist within Black cyberfeminism. Stemming from feminism's third wave, Black cyberfeminism represents a true engagement with the digital in the lives of wired women that encompasses a self-consciously critical stance toward the existing order with respect to the various ways that the digital affects women.

By bridging cyberfeminism and Black feminist thought, this framework is able to interrogate how women have understood their oppressed status, recognized the gendered and raced nature of the digital divide, and have made sense of their realities and experiences. Importantly, women are not passive bystanders in the information age waiting for their turn. As Collins (2000) confirmed, women have refused to become victims and have resisted marginalization in the information age. This resiliency has allowed women to bring the struggle to virtual spaces, thus empowering their communities.

Black women are urged to recognize the distinctness of our cultures. It is this deep heritage that provides us with the energy and skills needed to resist and transform daily discrimination. As women, we must embrace the history of our oppression, understanding that this history informs how power relations pervade our lives. We can't simply adopt privileged points of view and expect significant change to occur in either the virtual or physical world. We must embrace one another's oppressions and even privileges, recognizing that we all come from distinct realities converging in virtual spaces. There must be an affirmative action to be inclusive of a variety of women and viewpoints. Women working together is the only way to achieve significant changes. We cannot adopt the exclusionary approach of previous generations of women. We must recognize our privileges—racial, heterosexual, lingual, and so forth—and move toward fairness and equality for all women. Digital spaces provide us with a significant opportunity to accomplish this feat.

References

Brah, A., & Phoenix, A. (2013). Ain't I a woman? Revisiting intersectionality. *Journal of International Women's Studies, 5*(3), 75–86.

Brock, A., Kvasny, L., & Hales, K. (2010). Cultural appropriations of technical capital: Black women, weblogs, and the digital divide. *Information, Communication & Society, 13*(7), 1040–1059.

Bromseth, J., & Sundén, J. (2011). Queering internet studies: Intersections of gender and sexuality. In M. Consalvo & C. Ess (Eds.), *The handbook of internet studies* (pp. 270–299). London, UK: Wiley.

Brophy, J. E. (2010). Developing a corporeal cyberfeminism: Beyond cyberutopia. *New Media & Society, 12*(6), 929–945.

Chidester, D. (1992). *Religions of South Africa.* London, UK: Routledge.

Collins, P. H. (2000). *Black feminist thought: Knowledge, consciousness, and the politics of empowerment.* New York, NY: Routledge.

Crowley-Long, K. (1998). Making room for many feminisms. *Psychology of Women Quarterly, 22*(1), 113–130.

Daniels, J. (2009). Rethinking cyberfeminism(s): Race, gender, and embodiment. *WSQ: Women's Studies Quarterly, 37*(1), 101–124.

Daniels, J. (2013). Race and racism in Internet studies: A review and critique. *New Media & Society, 15*(5), 695–719.

Daniels, J. (2015). *The trouble with white feminism: Whiteness, digital feminism, and the intersectional internet* (Working paper). Retrieved February 21, 2015 from https://www.academia.edu/10851764/The_Trouble_with_White_Feminism_Whiteness_Digital_Feminism_and_the_Intersectional_Internet

Dei, G. J. S. (1995). Indigenous knowledge as an empowerment tool for sustainable development. In V. Titi & N. Singh (Eds.), *Empowerment for sustainable development: Toward operational strategies* (pp. 147–161). Winnipeg, Manitoba, Canada: Femwood.

Emerson, R. A. (2002). "Where my girls at?" Negotiating Black womanhood in music videos. *Gender & Society, 16,* 115–135.

Everett, A. (2004). On cyberfeminism and cyberwomanism: High-tech mediations of feminism's discontents. *Signs, 30*(1), 1278–1300.

Everett, A. (2007). *Learning race and ethnicity* (MacArthur Series, Digital Media and Learning). Boston, MA: The MIT Press.

Fesl, E. (1984). Eve Fesl. In R. Rowland (Ed.), *Women who do and women who don't join the women's movement* (pp. 109–115). London: Routledge and Kegan Paul.

Field, A. E., Cheung, L., Wolf, A. M., Herzog, D. B., Gortmaker, S. L., & Colditz, G. A. (1999). Exposure to mass media and weight concerns among girls. *Pediatrics, 103,* 54–60.

Florini, S. (2013). Tweets, tweeps, and signifyin': Communication and cultural performance on "Black Twitter." *Television & New Media, 15*(3), 223–237.

Glascock, J. (2001). Gender roles on prime-time network television: Demographics and behaviors. *Journal of Broadcasting & Electronic Media, 45,* 656–669.

Gray, K. L. (2012a). Deviant bodies, stigmatized identities, and racist acts: Examining the experiences of African-American gamers in Xbox Live. *New Review of Hypermedia and Multimedia, 18*(4), 261–276.

Gray, K. L. (2012b). Intersecting oppressions and online communities: Examining the experiences of women of color in Xbox Live. *Information, Communication & Society, 15*(3), 411–428.

Gray, K. L. (2013). Collective organizing, individual resistance, or asshole griefers? An ethnographic analysis of women of color in Xbox Live. *Ada: A Journal of Gender, New Media, and Technology, Issue 2.*Retrieved April 26, 2015 from http://adanewmedia.org/2013/06/issue2-gray/

Groesz, L. M., Levine, M. P., & Mumen, S. K. (2002). The effect of experimental presentation of thin media images on body satisfaction: A meta-analytic review. *International Journal of Eating Disorders, 31,* 1–16.

Hawthorne, S., & Klein, R. (Eds.). (1999). *Cyberfeminism: Connectivity, critique and creativity.* Melbourne, Australia: Spinifex Press.

Hobbs, P. (2004). In their own voices: Codeswitching and code choice in the print and online versions of an African-American women's magazine. *Women and Language, 27*(1), 1–12

hooks, b. (1992). *Black looks: Race and representation.* Boston, MA: South End Press.

Johnson, J. (2010). Manning up: Race, gender, and sexuality in Los Angeles' slam and spokenword poetry communities. *Text and Performance Quarterly, 30*(4), 396–419.

Kain, E. (2014, September 4). GamerGate: A closer look at the controversy sweeping video games. *Forbes.* Retrieved April 26, 2015 from http://www.forbes.com/sites/erikkain/2014/09/04/gamergate-a-closer-look-at-the-controversy-sweeping-video-games/

Kolko, B. (2000). Erasing @race: Going white in the (inter)face. In B. Kolko, L. Nakamura, & G. B. Rodman (Eds.), *Race in cyberspace* (pp. 213–232). New York, NY: Routledge.

Kolko, B., Nakamura, L., & Rodman, G. (Eds.). (2000). *Race in cyberspace.* New York, NY: Routledge.

Kopano, B. N. (2002). Rap music as an extension of the Black rhetorical tradition: "Keepin' it real." *The Western Journal of Black Studies, 26*(4), 204.

Kress, T. M. (2009). In the shadow of whiteness: (Re)exploring connections between history, enacted culture, and identity in a digital divide initiative. *Cultural Studies of Science Education, 4*(1), 41–49.

Levine, M. P., Smolak, L., & Hayden, H. (1994). The relation of sociocultural factors to eating attitudes and behaviors among middle school girls. *Journal of Early Adolescence, 74,* 471–490.

Lorde, A., & Clarke, C. (2007). *Sister outsider: Essays and speeches.* Berkeley, CA: Crossing Press.

Magnet, S. (2007). Feminist sexualities, race and the internet: An investigation of suicidegirls.com. *New Media & Society, 9*(4), 577–602.

Miller, J. B. (1991). Women and power. In J. V. Jordan, A. G. Kaplan, J. B. Miller, I. P. Stiver, & J. L. Surrey (Eds.), *Women's growth in connection: Writing from the Stone Center* (pp. 197–205). New York, NY: Guilford Press.

Minahan, S., & Cox, J. W. (2007). Stich'n bitch: Cyberfeminism, a third place and the new materiality. *Journal of Material Culture, 12*(1), 5–21.

Mouffe, C. (1992). Feminism, citizenship, and radical democratic politics. In J. Butler & J. W. Scott (Eds.), *Feminists theorize the political* (pp. 369–385). New York, NY: Routledge.

Narayan, U. (2013). *Dislocating cultures: Identities, traditions, and third world feminism.* New York, NY: Routledge.

Orgad, S. (2005). The transformative potential of online communication: The case of breast cancer patients' internet spaces. *Feminist Media Studies, 5*(2), 141–161.

Paasonen, S. (2002). Thinking through the cybernetic body: Popular cybernetics and feminism. *Rhizomes: Cultural studies in emerging knowledge, Issue 4.* Retrieved April 23, 2015 from http://www.rhizomes.net/issue4/paasonen.html

Paasonen, S. (2011). *Carnal resonance: Affect and online pornography.* Cambridge, MA: The MIT Press.

Patton, T. O. (2001). Ally McBeal and her homies: The reification of white stereotypes of the other. *Journal of Black Studies, 32,* 229–260.

Plant, S. (1997). *Zeros and ones: Digital women and the new technoculture.* London, UK: Fourth Estate.

Podlas, K. (2000). Mistresses of their domain: How female entrepreneurs in cyberporn are initiating a gender power shift. *CyberPsychology and Behavior, 3*(5), 847–854.

Potter, H. (2006). An argument for black feminist criminology: Understanding African American women's experiences with intimate partner abuse using an integrated approach. *Feminist Criminology, 1*(2), 106–124.

Puente S. (2008). From cyberfeminism to technofeminism: From an essentialist perspective to social cyberfeminism in certain feminist practices in Spain. *Women's Studies International Forum, 31*(6), 434–440.

Richardson, E. (2007). "She was workin' like foreal": Critical literacy and discourse practices of African American females in the age of hip hop. *Discourse and Society, 18*(6), 789–809.

Richardson, E. (2009). My ill literacy narrative: Growing up black, po and a girl, in the hood. *Gender and Education, 21*(6), 753–767.

Scott, K. D. (2000). Crossing cultural borers: 'Girl' and 'look' as markers of identity in Black women's language use. *Discourse and Society, 11,* 237–248.

Shih, J. (2006). Circumventing discrimination: Gender and ethnic strategies in Silicon Valley. *Gender and Society, 20*(2), 177–206.

Shirky, C. (2011). Political power of social media-technology, the public sphere and political change. *Foreign Affairs, 90*(1), 28–41.

Signorielli, N. (1997). *Reflections of girls in the media: A content analysis across six media.* Oakland and Menlo Park, CA: Children Now and Kaiser Family Foundation.

Smith, E. J. (1991). Ethnic identity development: Toward the development of a theory within the context of majority/minority status. *Journal of Counseling & Development, 70*(1), 181-188.

Stephens, D. P., & Few, A. L. (2007). The effects of images of African American women in hip hop on early adolescents' attitudes toward physical attractiveness and interpersonal relationships. *Sex Roles, 56*(3–4), 251–264.

Stephens, D. P., & Phillips, L. D. (2002*)*. Freaks, gold diggers, divas and dykes: The sociohistorical development of adolescent African-American women's sexual scripts *Sexuality and Culture, 7*(1), 3–49.

Wajcman, J. (1995). Feminist theories of technology. In S. Jasanoff, G. E. Markle, J. C, Petersen, & T. Pinch (Eds.), *Handbook of science and technology studies* (pp. 189–204). Thousand Oaks, CA: Sage.

Wajcman, J. (2004). *Technofeminism.* Cambridge, UK: Polity Press.

Wajcman, J. (2006). Technocapitalism meets technofeminism: Women and technology in a wireless world. *Labour & Industry, 16*(3), 7–20.

Wajcman, J. (2008). Life in the fast lane? Towards a sociology of technology and time. *The British Journal of Sociology, 59*(1), 59–77.

Ward, L. M., & Harrison, K. (2005). The impact of media use on girls' beliefs about gender roles, their bodies, and sexual relationships: A research synthesis. In E. Cole & J. H. Daniel (Eds.), *Featuring females: Feminist analyses of media* (pp. 3–23). Washington, DC: American Psychological Association.

CHAPTER TWELVE

Digital Human Rights Reporting by Civilian Witnesses: Surmounting the Verification Barrier

Ella McPherson

The scene is a dusty stretch—possibly of road—framed by rubble, old tires, barrels, abandoned vehicles, and crumbling walls. The footage is shaky, giving the impression that the camera is handheld. A man runs out of a doorway and shots ring out—small puffs of smoke erupting behind and ahead of him, suggesting that bullets are hitting the wall along which he runs. A moment of calm and then the camera pans left to a young boy, probably around eight years old, getting up from the ground. The boy begins to run toward an abandoned car; the shots recommence, and a puff of smoke emerges from his chest. He falls, slow-motion, first to his knees and then to his side. He lies there, face away from the camera, for a few seconds, then begins to run again, head down, toward the car. He drops behind it, then emerges dragging a younger girl in a bright pink top by the arm. They both run back the way he came, ducking at first, then running faster as the shooting continues. Throughout, male voices that seemingly issue from behind the lens excitedly talk, regularly crying *"Allah Akbar!"* The video ends a few seconds after the children exit the frame to the left.

This video first appeared on YouTube, posted by a new account on November 10, 2014, and titled "SYRIA! SYRIAN HERO BOY rescue girl in shootout." The video quickly went viral, reposted by Syrian activists and racking up millions of views. It also made headlines in the mass media; the *New York Post,* for example, ran a story the next day titled, "Harrowing Video Shows Boy Saving Girl from Sniper Fire in Syria" (Perez, 2014). Then, in a shocking turn of events—though not so shocking, perhaps, to the YouTube and reddit users who had been questioning the video's authenticity in comment threads—the BBC uncovered the video's cinematic origins

(Hamilton, 2014). Using funding from the Norwegian Film Institute and Arts Council Norway, a Norwegian director made the video in Malta on a set used in blockbuster films *Troy* and *Gladiator,* using professional child actors and Syrian refugees to provide the running commentary. According to the filmmakers, their funders were aware of their intent to post it online without the disclaimer that it was fiction. The filmmakers explained their motivations in a press release:

> By publishing a clip that could appear to be authentic we hoped to take advantage of a tool that's often used in war; make a video that claims to be real. We wanted to see if the film would get attention and spur debate, first and foremost about children and war. We also wanted to see how the media would respond to such a video. (Klevberg, 2014)

The backlash was tremendous. Social media users widely condemned the video, as did journalists and human rights workers. Several of these posted an open letter to the director and funders on a citizen journalism website. At the heart of their condemnation was the video's implication for the information documenting the Syrian conflict, namely that it "placed the burden of proof on those suffering rather than on those who cause the suffering." The letter explained:

> In such a conflict, deciphering the real from the fake is a difficult task and many activists, journalists and analysts spend countless hours sifting through videos in order to provide accurate information to the public. The intentionally misleading nature in which it was disseminated added to rather than detracted from the misinformation in Syria. This film will feed in to attempts to cast doubts on real stories coming out of Syria by citizen journalists and professional journalists alike. (Bellingcat, 2014)

The *Hero Boy* video highlights both the potential and the problem of produser documentation of facts for use by information professionals such as journalists and human rights researchers. On the one hand, if the video had been real, it would have been an example of how digital information and communication technologies (ICTs) facilitate new sources of information and new communication channels. Information and communication technology is a catchall category referring to the hardware and software that facilitate the production, storage, transmission, and reception of information (McKenzie, 2007); in this case, the ICTs of smartphones and social media are most relevant. In the context of a country such as Syria, largely closed to outside observers, YouTube videos are a crucial source of information for people within and without its borders and contribute to an information environment incomparable to the past. Consider, for example, the fact that local reports about the Syrian government's 2013 chemical weapons attack in Ghouta appeared on social media within hours. In contrast, the regime of President Hafez al-Assad was able to keep the 1982 Hama massacre under wraps for quite some time (Lynch, Freelon, & Aday, 2014). Facilitating the exercise of voice to report on atrocities, and specifically the exercise of civilian witnessing in the absence of professionals, is a potent manifestation of ICTs' pluralism potential.

On the other hand, the video turned out to be a fake and, as such, highlights the misinformation problem plaguing professionals trying to evaluate produser information for use in news, advocacy, and courts. In the Syrian case, instances of misinformation are easy to find. In some cases, misinformation even duped the experts, as when the BBC briefly used a photo of a child leaping over a row of shrouded bodies, which had been circulating on Twitter, to illustrate a 2012 Houla massacre. This photo was actually a photojournalist's image from 2003 in Iraq (Hamilton, 2012). The contrast between the social media-enabled potential of pluralism and the problem of misinformation has fueled a proliferation of tactics and technologies to support the verification of produser information. That said, verification remains a key barrier to the professional use of such information.

This chapter draws on the case of human rights fact finders' use of civilian witnesses' digital reports of human rights violations, although the issues it highlights are relevant for any professional use of produser information to establish facts. Using data from an ongoing digital ethnography of social media use in human rights work, I first explain why the verification of this information is so important yet can be so difficult. To do this, it is helpful to draw loosely on Bourdieu's (1983) classic sociological concept of a *field* of production. Participants in a field subscribe to a shared logic (or logics), namely the rules, explicit and implicit, that govern success in a particular field and thus the practices in that field (Thompson, 2010). Rules about the value and use of information are central to any field that trades in information, and we will see that verification is at the core of the human rights fact-finding field's information logic. In contrast, the digital produser category of civilian witnesses of human rights violations is largely a *non-field*. The meeting of this professional field and amateur non-field is where the verification barrier arises.

Verification of human rights information involves the corroboration of information's content and metadata (e.g., source, place, time, and conditions of production) using a variety of methods and sources; verification is necessary for the transformation of information into usable evidence. The verification of digital information is facilitated by *verification strategies* and *verification subsidies.* Verification strategies are part of the cultural capital, or knowledge central to success, of the human rights fact-finding field. A field's cultural capital can be spread through the field's networks, as seen in the verification knowledge exchange and training initiatives underway in the human rights fact-finding field (Bottero & Crossley, 2011). The idea of verification subsidies builds on Gandy's (1982) influential idea of information subsidies, namely tactics used by information producers to make it cheaper for others to use their information. Verification subsidies, powered by humans and machines, can either take on some of the labor required by various verification strategies or support the provision of metadata. Because accidental civilian witnesses usually are not members of a field, they often lack the networks necessary to build cultural capital about verification subsidies—or even to be aware of the need for verifiability. Given this, a number of third parties have innovated verification subsidies to lower the verification barrier between the human rights fact-finding field and the civilian witnessing non-field.

Verification as Core to the Information Logic of the Human Rights Fact-Finding Field

Fact-finding involves the gathering and evaluation of information, so it can be used as evidence in advocacy and in courts. Evidence is key to generating accountability for human rights violations (Clark, 2001). The accounts of human rights violations are often contested, however, by the accused on two fronts: the veracity of the evidence and the credibility of its source. A human rights organization's reputation for credibility is a fundamental asset, not only in advocacy but also for garnering donations and volunteers, motivating mobilization, and influencing policy making. To protect the integrity of evidence as well as institutional reputation, the human rights fact-finding field has developed robust and transparent fact-finding methodologies that emphasize verification strategies (Brown, 2008; Edwards & Koettl, 2011; Gibelman & Gelman, 2004; Hopgood, 2006; Land, 2009a, 2009b; Orentlicher, 1990; Satterthwaite & Simeone, 2014). For example, Physicians for Human Rights' (2014) digital *Map of Attacks on Health Care in Syria* links to a more than 3,000-word exposition of its methodology. Human Rights Watch (2014) similarly features a lengthy description of its research methodology on the publications section of its website.

Because of the expertise that fact-finding requires, the development of human rights methodologies has evolved in tandem with the professionalization of human rights (Alston, 2013; Land, 2009b). Intergovernmental organizations drove a first generation of fact-finding, in which diplomats, experts, and lawyers reviewed on-the-ground research to write reports for governments and intergovernmental groups. The large, international, human rights nongovernmental organizations (NGOs) drove a second generation, which drew largely on witness interviews and produced reports targeted at public opinion as well as political bodies. The third (current) generation is characterized by flexibility in fact-finding methodology and research output. This generation is born of ICTs and a growing number and diversity of contributors to human rights fact-finding—and, in particular, the digitally enabled escalation of information produced and transmitted by civilian witnesses (Alston, 2013; Satterthwaite, 2013).

The *Non-Field* of Digital Civilian Witnessing of Human Rights Violations

At the risk of further proliferating descriptors about produsers, I use the concept of *civilian witnesses* to refer to eyewitnesses digitally documenting human rights violations (Allan, 2013). Through the definition of a civilian as outside of the profession in question, this nomenclature highlights the inexpert nature of the production of information by civilian witnesses.[1] Information from civilian witnesses has long been a cornerstone of human rights fact-finding, in part because of the human rights community's commitment to amplifying the voices of those holding powerful actors to account (Satterthwaite, 2013; Satterthwaite & Simeone, 2014). Traditionally, this information tended to be gathered by human rights fact finders conducting on-the-ground research or witness interviews. As such, fact finders had oversight of the production and gathering of information. The introduction of ICTs such as smartphones and social media platforms has facilitated civilian witnesses' autonomous production and transmission of information. In other words, ICTs support the rise of amateurs in a fact-finding process traditionally dominated by professionals (Land, 2009b).

That said, people spontaneously reporting on human rights violations via ICTs are a varied bunch—and not entirely unprofessional. At one end of the spectrum are activists who can be thought of as belonging to what Postill (2015) called non-institutionalized movement-fields. These fields are characterized by logics and networks facilitating the sharing of cultural capital—including tips traded by activists on how to produce verifiable information, such as stating the time and place while filming, providing a shot of a newspaper, and panning the horizon for landmarks (Wardle, 2014). But this more professionalized end of the civilian witness spectrum is akin to what Human Rights Watch (2014) called the "trusted contacts" whom human rights fact finders traditionally relied on for information. What is new and of concern here are the number and variety of *accidental* civilian witnesses reporting information digitally at the other end of the civilian witness spectrum. Accidental civilian witnesses are those who happen to be, as Murphy (2013) put it, "in the wrong place at the wrong time" and who choose to document and share unfolding events. This category of civilian witness is a non-field; people engage in the practice of human rights reporting randomly or sporadically and by definition have neither a shared logic nor—in all likelihood—cultural capital in the form of digital verification literacy (or the networks to gain it).

The affordances of smartphones and social media underpin the proliferation of these accidental civilian witnesses. Affordances are what new technologies allow their users to do and are shaped by the technology's characteristics or materialities (Treem & Leonardi, 2012). A key affordance here is the facilitation of user-generated content through mobile devices equipped with cameras that support the capture of information and through social media platforms that ease its transmission. Another is the disembodiment of information from the time, place, and source of production—which means information can, for example, cross closed borders more easily than ever before (McPherson, 2014).

The very affordances that allow for information's production and transmission by civilian witnesses can be hindrances for human rights fact finders who are gathering and evaluating information. The user-generated content affordance has created a flood of information from unknown sources puporting to document human rights violations all over the world. Collecting and organizing this information for evaluation poses the risk of overwhelming the fact-finding teams at human rights organizations. As Lara Setrakian, editor-in-chief of online news outlet *Syria Deeply* put it, ". . . The Syria story has a big data problem" (Setrakian, 2013). Additionally, the disembodiment affordance means that metadata, which are so crucial to the evalution of the information's veracity, may be meager or missing. This may be because civilian witnesses may not know to supply metadata or because many mainstream social media platforms strip out this metadata during upload. As such, the digitally enabled rise of civilian witnesses means that human rights fact finders are faced with more information that is more difficult to verify than ever before.

Verification Strategies in the *Field* of Professional Human Rights Fact-Finding

Although the nature of information being verified and the individual techniques for verification are shifting rapidly as ICTs evolve, the "fundamentals of verification" remain constant (Silverman & Tsubaki, 2014, p. 8). Fact finders must establish the source, place, and time of production and cross-reference content and metadata with other methods and sources (Wardle, 2014). Verification strategies under the dual pressures of the digital information deluge and the limited resources of human rights organizations and civilian witnesses are about securing and speeding this verification process as much as possible and may be facilitated by the relatively young field of digital *information forensics:* the use of digital tools and databases for verification.

Identifying the original source of the information requires establishing its chain of custody. This process can unearth instances where information has been manipulated, as in the scraping practice prevalent on YouTube, in which one user reuploads another's YouTube content with no indication of its provenance (Wardle, 2014). Reverse image search platforms such as those provided by Google or TinEye can help; users can upload or link to an image, and these platforms will return the locations of matching copies.

Once the original source is identified, the fact finder should ideally speak with the individual to get his or her story of witnessing and documenting the event (Barot, 2014). Verification experts recommend that the fact finder then examine the digital footprint of the source, looking at the individual's organizational affiliations, posting history, followers and friends, and location (Browne, 2014; Kilroy, 2013; Meier, 2011a; Silverman & Tsubaki, 2014). For example, the longevity of a social media account may help indicate credibility, because users who post misinformation have been known to do it via a new account (Koettl, 2014a). "There is no quick way" of verifying the identity of a social media account, Wardle (2014) warned; rather, source verification requires "painstaking checks," akin to "old-fashioned police investigation" (pp. 29–30).

The next stage is identifying and corroborating time and place of production. The civilian witness may divulge this information in an interview or may have included it with the file. The latter can occur either in the production process, such as by stating the location and date during filming, or in the transmission process, such as by providing commentary upon upload to YouTube. Time and place may be evident from landmarks, shadows, weather, signage, clothing, weapons, and dialect captured in the digital file

(Kilroy, 2013; Koettl, 2014a). They may also be identified via metadata automatically embedded in the file; digital platforms such as FotoForensics and the Citizen Evidence Lab's YouTube Data Viewer facilitate metadata extraction (Barot, 2014; Koettl, 2014c). Content and metadata can be corroborated via digital databases, which can easily be searched for digital footprints (i.e., pipl.com), landmarks (Google Maps), and weather (Wolfram Alpha), for example (Silverman, 2014). It may be cross-referenced against other digital files, such as by time-syncing multiperspective produser videos of the same event or by comparing social media information with satellite images (The Rashomon Project, n.d.; Wang, Raymond, Gould, & Baker, 2013). Another source of digital data for cross-reference is satellite images. Of course, digital content and metadata can also be corroborated offline through on-the-ground research and consulting trusted and expert networks, including forensic pathologists to assess the causes of digitally documented deaths (McPherson, 2014).

Building Cultural Capital About Digital Verification

The successful verification of digital information benefits from the dissemination of cultural capital among fact finders and civilian witnesses. Specifically, fact finders should learn about digital verification strategies, and civilian witnesses should learn about verification subsidies. Various initiatives are underway to disseminate this cultural capital, although their spread depends in part on the extent to which their target audiences are networked.

For fact finders, examples include the *Verification Handbook,* published in early 2014 by the European Journalism Centre, which gathers insights from leading verification experts (Silverman, 2014). Amnesty International's Citizen Evidence Lab website, also launched in 2014, provides guidance as well as a knowledge exchange space for human rights fact finders about using digital information (Koettl, 2014a). Human rights organization New Tactics in Human Rights (2014) hosted an online conversation, publicly available and archived, on "Using Video for Documentation and Evidence."

For civilian witnesses, Witness provides a guide on including key information in human rights videos—such as the what, when, where, and (if safe) the who of the video. The guide explains:

> Adding this information to videos will make it much easier for reviewers that were not at the scene of the human rights incident to verify the content. Easier verification

means there is a better chance that the video will be used to secure justice. (Witness, n.d.)

Verification training initiatives must come to the attention of fact finders and civilian witnesses in order for them to build their cultural capital. As evident in the open knowledge exchange events run by New Tactics in Human Rights, the human rights field is characterized by networked solidarity that supports the diffusion of verification strategies (Atack, 1999; Dütting & Sogge, 2010; Keck & Sikkink, 1998; McLagan, 2006). At the activist end of the civilian witness spectrum, networked solidarity may also be enabling the diffusion of knowledge about verification subsidies. One of my human rights fact finder interviewees told me the following, in reference to Syria:

> And I have to say there has been an improvement in the way that activists are putting the information online. At the beginning, they were thinking that just by taking this video online, it could be useful. But with practice, they have begun to learn: No, you need to record things. You need to say who this person is or where this is happening. By whom.

In contrast, the non-field of accidental civilian witnessing does not provide natural opportunities for learning either about the informational needs of the verification process or about how civilian witnesses might support these with verification subsidies. An alternative to the difficult task of building accidental civilian witnesses' cultural capital is the provision of verification subsidies by third parties.

Verification Subsidies

Verification subsidies tend to fall into one of two categories: those that supplant some of the labor of fact finders and those that enhance the information provided by civilian witnesses. In each category, both human-driven and machine-driven initiatives exist. The deployment of third-party verification subsidies involves a variety of actors—from NGOs and the academy to commercial actors making a business of verification (such as Storyful) and those that are subject to lobbying by NGOs to build verification subsidies into their ICTs.

In terms of supplanting fact finders' evaluation of digital information from civilian witnesses, a human-powered option is to harness the crowd. Crowdsourcing involves institutions turning over tasks traditionally done by a specific individual to a big, unspecified group recruited through an open call—although, in bounded crowdsourcing, crowd membership is limited by invitation (Howe, 2006; Meier, 2011b). One example is Veri.ly, a web-based project in development via collaboration between the University of Southampton, Masdar Institute, and the Qatar Computing Research Institute. Veri.ly asks contributors to supply yes-or-no answers to questions about veracity supported with textual or visual evidence (Naroditskiy, 2014; Veri.ly, 2014). The site deliberately asks for supporting documentation because "verification . . . requires searching for evidence rather than just liking or retweeting something," in the words of co-inventor Victor Naroditskiy (2014). As such, Naroditskiy explained that Veri.ly is deliberately positioned in contrast to sites such as reddit, which allows up or down votes and notoriously facilitated the crowd's misidentification of the Boston Marathon bombers. An example of a bounded crowd is the Citizen Media Evidence Partnership, a project being developed by Amnesty International's Sensor Project and Will H. Moore of Florida State University. This project trains student groups to become "verification corps" dedicated to triaging information for verification by experts (Koettl, 2014b; Moore, 2014).

Others are exploring the machine provision of verification subsidies for fact finders. TweetCred, for example, developed by a team of academics from the Indraprastha Institute of Information Technology and the Qatar Computing Research Institute, attempts this through a machine-learning algorithm. TweetCred rates the credibility of individual tweets in real time by evaluating them against 45 criteria. These include aspects of the tweet's content, such as emoticons, swearwords, and hashtags, and the metadata of the tweet and its author, such as number of followers (Gupta, Kumaraguru, Castillo, & Meier, 2014).

In terms of the second category of verification subsidies, those that address the deficit in metadata in information supplied by civilian witnesses, one approach is to encourage civilian witnesses to include metadata at the point of transmission. The Syria Tracker project, for example, which crowdsources information about the conflict in Syria, requires submitters to enter a title, description, category (including "Killed," "Missing," "Revenge Killings," and "Eyewitness Report"), their location, and the time of reporting. Optional information includes digital images, contact information, and links to news sources and external video (Syria Tracker, n.d.).

Machine-supported options to enhance civilian witness information automate the inclusion of metadata at the point of capture. InformaCam, under development by Witness and the Guardian Project, allows users to embed data about the time, date, and place of videos or photos via information gathered by their smartphones' sensors from cell towers, Wi-Fi, and Bluetooth. Users can employ digital signatures and encryption and can transmit files securely over Tor; the aim is for InformaCam to allow for the collection and transmission of "ironclad digital media that can be used in courts of international law," according to Sam Gregory of Witness (InformaCam, n.d.).

ICTs that engage civilian witnesses in the provision of verification subsidies have the added benefit of training them in how to produce and transmit future reports. Of course, this depends on civilian witnesses knowing about and using these platforms rather than mainstream social media platforms, which do not, at the moment, prompt for or automate comprehensive metadata inclusion. Witness is, however, advocating for the inclusion of an "eyewitness mode" resembling InformaCam as standard in preloaded photo and video applications on smartphones and in social media platforms (Gregory, 2014). The inclusion of this feature in mainstream applications and platforms means that accidental civilian witnesses may gain cultural capital about verification by simply exploring standard features of their smartphones and social media accounts. Even so, the accidental civilian witness will not need much in the way of cultural capital concerning verification to provide subsidies through the eyewitness mode.

Pluralism and Power at the Boundary of the *Field* of Professional Human Rights Fact-Finding and the *Non-Field* of Digital Civilian Witnessing

In sum, what I have shown is that a barrier exists between the field of human rights fact-finding and the non-field of accidental civilian witnessing: verification. To help understand why this barrier exists, I have loosely relied on Bourdieu's (1983) concept of field. This concept is analytically useful for at least four reasons. First, it allows us to imagine the social space of human rights fact-finding as delimited by a boundary. Gaining access to this space—and for those documenting a human rights violation, access to this space is a key channel for generating accountability—requires the surmounting of particular barriers. Second, for fields that trade in information, such as human rights fact-finding, rules about the value and use of information are central to their logics. These information values and uses can form some of the barriers to the field; in the case of human rights fact finders, informational veracity is highly valued, both for its own sake in terms of the transformation of information into evidence for use in advocacy and courts and because of its implications for the symbolic capital of credibility so vital to human rights organizations. These two characteristics of fields help us begin to understand the meeting point of produser information and professional fact finders. They highlight that, to understand which produsers' information gets used by professional fact finders, we need to understand both the information values that are part of the professional field's logic and the ability of the amateur produsers to meet these values— namely, the ability to provide what Gandy (1982) called information subsidies, or, in our case, verification subsidies.

A third reason the concept of field is useful is because it allows us to contrast it with the non-field. Especially at the accidental end of the activist-accidental civilian witness spectrum, those providing digital information are a non-field. As Bottero and Crossley (2011) posited—and the fourth reason using the concept of field is useful for understanding the professional-produser information interface—fields can be thought of as containing networks that allow for the diffusion of cultural capital; the human rights fact-finding field, for example, has resources and institutions allowing for knowledge exchange about verification strategies. In contrast, the absence of field means the absence of this type of network, so it is difficult for cultural capital about verification subsidies to spread among accidental civilian witnesses. In the context of the volume problem of social media information and the verification literacy problem of digital civilian witnesses, verification subsidies, many provided by third parties, can support fact finders and civilian witnesses with human and machine endeavors.

This lay of the land brings us back to the juxtaposition raised by the Syria *Hero Boy* video: the problem of verification versus the promise of pluralism. Pluralism, like veracity, is a core value in the human rights logic. Speaking truth to power is a major aim of human rights work, and this entails listening to the powerless, namely those otherwise likely to be drowned out by the actors they accuse of violations (Satterthwaite, 2013). The implications of digitally enabled produsers for human rights pluralism are complex. On the one hand, ICTs have undoubtedly increased the pluralism of human rights information, especially from closed contexts such as Syria. On the other, this increase in the pluralism of information does not necessarily correspond with an increase in the pluralism of evidence. The same technological affordances making it easier for produsers to provide information can make professional verification more difficult. Because information is only acted on once it becomes evidence, the volume and variety of civilian witnesses able to provide verification subsidies may be connected to the pluralism of access to the accountability mechanism of human rights (McPherson, 2014).

One of the most analytically useful aspects of Gandy's information subsidy concept is that, as an economic metaphor, it highlights the connection between the ability to provide information subsidies and the possession of other forms of capital. The ability to provide verification subsidies relates to cultural capital in the form of digital literacy, as well as symbolic capital in the form of a digital footprint—which allows fact finders to more easily corroborate the metadata of source identity. This is potentially problematic for at least two reasons. First, after Bourdieu, can we assume that these forms of capital correspond to power and correlate to other forms of capital, such that the less powerful have more trouble producing verifiable information? Second, although the ability to provide verification subsidies can determine whether something is verifiable, and thus usable, it also determines ease of verification. Gandy (1982) posited that information consumers, in the context of finite resources, are more likely to consume cheaper information; will verification ease determine the choice of human rights information in the context of limited time and manpower and a deluge of digital information? If the answers to these questions are yes, more powerful civilian witnesses may have greater access to the mechanisms of human rights accountability—despite the fact that it may be the less powerful civilian witnesses who have greater need.

This is where third-party verification subsidies hold promise. Through supplanting some of the fact finders' verification labor, and by prompting and automating the provision of metadata in civilian witness information, they make it easier for fact finders to evaluate more digital information for evidence. Interestingly, the most promising verification subsidy may be the technological solution of the eyewitness mode in smartphone and social media platforms. Through providing verification technology and tips at produsers' fingertips, this subsidy can mitigate the non-field problem of limited diffusion of cultural capital. In this case, then, automation, by taking human activity out of the process, may bring more humans into the product and thus potentially increases the pluralism of human rights evidence. This verification subsidy works, however, because it replaces some of the technical work of verification. Machine verification subsidies, such as Tweet-Cred, that try to address the subjective aspects of verification, are more dubious. Determining the presence of a fact or the nature of truth is a subjective judgment and thus, at least for the foreseeable future, a human process.

Finally, this analysis of where the amateur non-field of civilian witnesses meets the professional field of human rights fact finders is useful for reflecting on field theory. As Lahire (2015) argued, focusing on fields—that is, journalism, art, politics, human rights—as sites of study creates a focus on professional domains and thus on the relatively elite and powerful individuals who participate in these social spaces. The rise of the digitally enabled produser, so often an amateur and thus in a non-field, has rightfully made us refocus our attention—as Lahire urged we do—on the study of these less-powerful actors and on what happens when they interact with professional fields. Just as Bourdieu was so concerned with power relations in and across fields, our understanding of human rights civilian witnesses' differing abilities to provide verification subsidies impels us also to recognize and be concerned with power relations in non-fields.

Notes

This work was supported by the Economic and Social Research Council (grant number ES/K009850/1) and by the Isaac Newton Trust.

1. "Civilian" is preferable to "citizen," which is oft used to refer to produsers but implies a relationship with the state—problematic in the case of those accusing this state of violating human rights. Furthermore, "witness" is also preferable to "journalist," as it both highlights the amateur and accidental nature of this activity as well as the intentionality of engendering a response (Allan, 2013; Joyce, 2013).

References

Allan, S. (2013). *Citizen witnessing: Revisioning journalism in times of crisis.* Cambridge, UK: Polity Press.

Alston, P. (2013). Introduction: Third generation human rights fact-finding. *Proceedings of the Annual Meeting (American Society of International Law)*, 107, 61–62.

Atack, I. (1999). Four criteria of development NGO legitimacy. *World Development, 27*(5), 855–864.

Barot, T. (2014). Verifying images. In C. Silverman (Ed.), *Verification handbook: A definitive guide to verifying digital content for emergency coverage* (pp. 33–39). Maastricht, the Netherlands: European Journalism Centre.

Bellingcat. (2014, November 17). *An open letter to Lars Klevberg, The Norwegian Film Institute and Arts Council Norway.* Retrieved December 10, 2014 from https://www.bellingcat.com/news/uk-and-europe/2014/11/17/an-open-letter-to-lars-klevberg-the-norwegian-film-institute-and-arts-council-norway/

Bottero, W., & Crossley, N. (2011). Worlds, fields and networks: Becker, Bourdieu and the structures of social relations. *Cultural Sociology, 5*(1), 99–119.

Bourdieu, P. (1983). The field of cultural production, or: The economic world reversed. *Poetics, 12*(4–5), 311–356.

Brown, L. D. (2008). *Creating credibility*. Sterling, VA: Kumarian Press.

Browne, M. (2014). Verifying video. In C. Silverman (Ed.), *Verification handbook: A definitive guide to verifying digital content for emergency coverage* (pp. 45–51). Maastricht, the Netherlands: European Journalism Centre.

Clark, A. M. (2001). *Diplomacy of conscience: Amnesty International and changing human rights norms*. Princeton, NJ: Princeton University Press.

Dütting, G., & Sogge, D. (2010). Building safety nets in the global politic: NGO collaboration for solidarity and sustainability. *Development, 53*(3), 350–355.

Edwards, S., & Koettl, C. (2011, June 20). *The Amnesty crowd: Mapping Saudi Arabia*. Retrieved March 3, 2015 from http://www.ushahidi.com/2011/06/20/amnesty-crowd/

Gandy, O. H. (1982). *Beyond agenda setting: Information subsidies and public policy*. Norwood, NJ: Ablex.

Gibelman, M., & Gelman, S. R. (2004). A loss of credibility: Patterns of wrongdoing among nongovernmental organizations. *Voluntas: International Journal of Voluntary and Nonprofit Organizations, 15*(4), 355–381.

Gregory, S. (2014, March 3). *How an eyewitness mode helps activists (and others) be trusted*. Retrieved March 3, 2015 from http://blog.witness.org/2014/03/eyewitness-mode-helps-activists/

Gupta, A., Kumaraguru, P., Castillo, C., & Meier, P. (2014). *TweetCred: A real-time web-based system for assessing credibility of content on Twitter*. Retrieved March 3, 2015 from http://arxiv.org/abs/1405.5490

Hamilton, C. (2012, May 29). Houla Massacre picture mistake. Retrieved from http://www.bbc.co.uk/blogs/legacy/theeditors/2012/05/houla_massacre_picture_mistake.html

Hamilton, C. (2014, November 18). *How we discovered the truth about YouTube's Syrian "hero boy" video*. Retrieved March 3, 2015 from http://www.bbc.co.uk/blogs/blogcollegeofjournalism/posts/How-we-discovered-the-truth-about-YouTubes-Syrian-hero-boy-video

Hopgood, S. (2006). *Keepers of the flame: Understanding Amnesty International*. Ithaca, NY: Cornell University Press.

Howe, J. (2006, June 2). *Crowdsourcing: A definition*. Retrieved March 3, 2015 from http://crowdsourcing.typepad.com/cs/2006/06/crowdsourcing_a.html

Human Rights Watch. (2014). *Our research methodology*. Retrieved March 3, 2015 from http://www.hrw.org/node/75141

InformaCam. (n.d.). *InformaCam—Secure & verified mobile media*. Retrieved March 3, 2015 from https://guardianproject.info/informa/

Joyce, D. (2013). Media witnesses: Human rights in an age of digital media. *Intercultural Human Rights Law Review, 8*(232), 231–280.

Keck, M. E., & Sikkink, K. (1998). *Activists beyond borders: Advocacy networks in international politics*. Ithaca, NY: Cornell University Press.

Kilroy, D. (2013, February 25). *Citizen video for journalists: Verification*. Retrieved March 3, 2015 from http://blog.witness.org/2013/02/citizen-video-for-journalists-verification/

Klevberg, L. (2014). Twitter photo [microblog]. Retrieved March 3, 2015 from https://twitter.com/LarsKlevberg/status/533385072083501056/photo/1

Koettl, C. (2014a). *Human rights citizen video assessment tool.* Retrieved March 3, 2015 from http://fluidsurveys.com/s/citizenvideo-stress-test/

Koettl, C. (2014b, February 18). *"The YouTube war": Citizen videos revolutionize human rights monitoring in Syria.* Retrieved March 3, 2015 from http://www.pbs.org/mediashift/2014/02/the-youtube-war-citizen-videos-revolutionize-human-rights-monitoring-in-syria/

Koettl, C. (2014c, July 1). *YouTube data viewer.* Retrieved March 3, 2015 from http://citizen evidence.org/2014/07/01/youtube-dataviewer/

Lahire, B. (2015). The limits of the field: Elements for a theory of the social differentiation of activities. In M. Hilgers & E. Mangez (Eds.), *Bourdieu's theory of social fields: Concepts and applications* (pp. 62–101). Abingdon, UK: Routledge.

Land, M. B. (2009a). Networked activism. *Harvard Human Rights Journal, 22,* 205–243.

Land, M. B. (2009b). Peer producing human rights. *Alberta Law Review, 46*(4), 1115–1139.

Lynch, M., Freelon, D., & Aday, S. (2014). *Syria's socially mediated civil war* (Peaceworks No. 91). Washington, DC: United States Institute of Peace. Retrieved March 3, 2015 from http://www.usip.org/publications/syria-s-socially-mediated-civil-war

McLagan, M. (2006). Introduction: Making human rights claims public. *American Anthropologist, 108*(1), 191–195.

McKenzie, D. J. (2007). *Youth, ICTs, and development.* Washington, DC: World Bank. Retrieved March 5, 2015 from http://documents.worldbank.org/curated/en/2007/03/7458887/youth-icts-development

McPherson, E. (2014). Advocacy organizations' evaluation of social media information for NGO journalism: The evidence and engagement models. *American Behavioral Scientist, 59*(1), 124–148.

Meier, P. (2011a). *How to verify social media content: Some tips and tricks on information forensics.* Retrieved March 3, 2015 from http://irevolution.net/2011/06/21/information-forensics/

Meier, P. (2011b, December 7). *Why bounded crowdsourcing is important for crisis mapping and beyond.* Retrieved March 3, 2015 from http://irevolution.net/2011/12/07/why-bounded-crowdsourcing/

Moore, W. (2014, May 21). *About.* Retrieved March 3, 2015 from http://c-mep.org/home/about/

Murphy, H. (2013, March 18). A guide to watching Syria's war [Video file]. *The New York Times.* Retrieved March 3, 2015 from http://www.nytimes.com/video/world/middleeast/100000002124826/watching-syrias-war.html

Naroditskiy, V. (2014, August 13). *Veri.ly—Getting the facts straight during humanitarian disasters.* Retrieved March 3, 2015 from http://www.software.ac.uk/blog/2014-08-13-verily-getting-facts-straight-during-humanitarian-disasters

New Tactics in Human Rights. (2014). *Using video for documentation and evidence.* Retrieved March 3, 2015 from https://www.newtactics.org/conversation/using-video-documentation-and-evidence

Orentlicher, D. F. (1990). Bearing witness: The art and science of human rights fact-finding. *Harvard Human Rights Journal, 3,* 83–136.

Perez, C. (2014, November 11). *Harrowing video shows boy saving girl from sniper fire in Syria.* Retrieved March 3, 2015 from http://nypost.com/2014/11/11/brave-boy-appears-to-save-girl-from-sniper-fire-in-syria/

Physicians for Human Rights. (2014). *Anatomy of a crisis, a map of attacks on health care in Syria: Methodology.* Retrieved March 3, 2015 from https://s3.amazonaws.com/PHR_syria_map/methodology.pdf

Postill, J. (2015, January 16). *Six ways of doing digital ethnography.* Retrieved March 3, 2015 from http://johnpostill.com/2015/01/16/13-six-ways-of-researching-new-social-worlds/

Satterthwaite, M. (2013). Finding, verifying, and curating human rights facts. *Proceedings of the Annual Meeting (American Society of International Law),* 107, 62–65.

Satterthwaite, M. L., & Simeone, J. (2014). *An emerging fact-finding discipline? A conceptual roadmap for social science methods in human rights advocacy* (New York School of Law, Public Law Research Paper No. 14-33). New York: New York University School of Law. Retrieved March 3, 2015 from http://papers.ssrn.com/abstract=2468261

Setrakian, L. (2013, April 1). *Citizen video for journalists: Contextualization.* Retrieved March 3, 2015 from http://blog.witness.org/2013/04/citizen-video-for-journalists-contextualization/

Silverman, C. (Ed.). (2014). *Verification handbook: A definitive guide to verifying digital content for emergency coverage.* Maastricht, the Netherlands: European Journalism Centre.

Silverman, C., & Tsubaki, R. (2014). When emergency news breaks. In C. Silverman (Ed.), *Verification handbook: A definitive guide to verifying digital content for emergency coverage* (pp. 5–12). Maastricht, the Netherlands: European Journalism Centre.

Syria Tracker. (n.d.). *Submit a report.* Retrieved March 3, 2015 from https://syriatracker.crowdmap.com/reports/submit

The Rashomon Project. (n.d.). *About Rashomon.* Retrieved from http://rieff.ieor.berkeley.edu/rashomon/about-rashomon/

Thompson, J. B. (2010). *Merchants of culture.* Cambridge, UK: Polity Press.

Treem, J. W., & Leonardi, P. M. (2012). Social media use in organizations: Exploring the affordances of visibility, editability, persistence, and association. *Communication Yearbook,* 36, 143–189.

Veri.ly. (2014, October 21). *Is this Highgate Cemetery?* Retrieved January 11, 2015, from https://veri.ly/crisis/3/question/81-is-this-highgate-cemetery

Wang, B. Y., Raymond, N. A., Gould, G., & Baker, I. (2013). Problems from hell, solution in the heavens?: Identifying obstacles and opportunities for employing geospatial technologies to document and mitigate mass atrocities. *Stability: International Journal of Security & Development,* 2(3), 53.

Wardle, C. (2014). Verifying user-generated content. In C. Silverman (Ed.), *Verification handbook: A definitive guide to verifying digital content for emergency coverage* (pp. 24–33). Maastricht, the Netherlands: European Journalism Centre.

Witness. (n.d.). *Video as evidence: Adding essential information to video.* Retrieved March 3, 2015 from http://library.witness.org/product/video-evidence-adding-essential-information-video/

Twitter as a Pedagogical Tool in Higher Education

Renee Hobbs

When my students first learn that they will have to use Twitter as part of the learning experience, a few are enthusiastic, some are bemused, and some are downright hostile. In resisting my request for students to establish a Twitter account, one student said to me, angrily, "I don't want to have to talk about Justin Bieber's haircuts!"

In 2008, I joined Twitter as @reneehobbs and sent my first tweet. Since then, I have been experimenting with Twitter as a tool for personal growth and have used it for community outreach, social marketing, research, and teaching. It's not an understatement to say that Twitter has become just as indispensable as *The New York Times* for my own monitoring and surveillance of the social world. Although e-mail continues to be a valuable tool for maintaining social relationships with colleagues, Twitter's ease, information value, and reach in helping me connect to new networks of scholars and professionals helps me keep up with the ever-rising tide of research, information, opinions, and ideas. Certain individuals with expertise in education, media, technology, and information policy, including @AudryWatters and @MatthewIngram, have become vital and respected thought leaders to me; reading their tweets and learning from the resources they share are part of my professional learning routine. Among my friends and colleagues, Twitter gives me a sense of connection to their daily lives, a feeling that can be surprisingly intimate. Most importantly, interacting with people on Twitter has affected my ongoing understanding of media, culture, and technology industries in education. So naturally, I have become interested in exploring how to introduce learners to Twitter so they may benefit from it, too.

In this chapter, I explore how the use of Twitter can be valuable particularly in the context of online learning in graduate higher education. As digital media and technologies become essential dimensions of learning and teaching, I'm interested in how educators activate students' learning progression to support the use of Twitter as an information and communication tool. In what follows, I describe some of the ways that Twitter may support the learning process in higher education, especially in the context of online learning in the graduate education of communication professionals and future school and public librarians. First, however, I contextualize this topic in relation to the theoretical frames that position digital media and technologies as essential dimensions of learning and teaching, considering the rise of online learning in professional programs. Then I share examples of student tweets to illustrate some tentative and provisional ideas about the affordances of Twitter as a teaching tool and the learning progression involved in using Twitter as a learning resource. Finally, I consider some of the gaps and omissions that are evident from an examination of my initial experimentation with Twitter, which provide opportunities for me to reflect on practices that support the process of producing tweets as a form of learning.

Beyond the Walled Garden

Educators at all levels have made effective use of media and communication texts, tools, and technologies in both formal and informal contexts (Cuban, 1993; Hobbs, 2010; Jenkins, Purushotma, Clinton, Weigel, & Robinson, 2006; Mollett, Moran, & Dunleavy, 2011). Although social media such as Facebook and Twitter have been widely adopted by students in their personal lives, research on the use of social media in the context of education is still in its infancy. We know little about how serendipitous, surprising, and unexpected dimensions of information exploration and social interaction can lead to meaningful learning (Buchem, 2012). From personal observation, I know that browsing for short updates can be a powerful experience, albeit quite different from using a search engine such as Google. Twitter browsing can be similar to the practice of wandering the stacks of the university library: Using Twitter, I have stumbled on many valuable resources by happenstance. The pleasurable sense of discovering something unexpected can invigorate my intellectual curiosity and open my thinking in new directions. Moreover, as Thompson (2007) observed, Twitter's constant-contact sensibility creates a form of social proprioception, where granular updates from colleagues may increase our awareness of each other's thoughts, ideas, emotions, and experiences. Often the material shared on Twitter is hot off the press, and because the people I follow on Twitter come from varied business and academic sectors, the materials they choose to share come from discourse communities in government, the humanities, engineering, business, nonprofit and social service sectors, and civic activism. Through Twitter, I have widened my social network while discovering highly relevant information and participating in social learning experiences.

The rise in interest in the use of digital and social media tools for learning in formal contexts such as higher education is in part a response to the changing economics of higher education. Consider the impossible economics of higher education. Student loan debt has risen 20% to $1.2 trillion between 2011 and 2013 and now exceeds every other form of nonmortgage debt. The debt load of the average college graduate is about $30,000 (Berr, 2014). The ripple effects of this debt extend in predictable but distressing ways: Those with high levels of student debt are less likely to start new businesses, buy homes, or seek careers in public service (Korrki, 2014). This evidence will not surprise many privileged professors, including those reading this essay, but we must recognize our culpability in maintaining status quo realities that are having a serious negative social impact on the next generation. For this reason, robust experimentation with online learning is essential as we discover new ways to meet the needs of learners in an educationally robust yet cost-effective way.

Online learning forces learners to assume a highly active position with great levels of personal responsibility and time management. Students move from a culture of dependency on the teacher to a culture of autonomy because they create, share, manage, and collaborate on their learning with other learners (Tu, 2014). For educators who are responsible for promoting students' multimedia authorship competencies, online learning has a number of advantages because it enables students to combine written and spoken language, sound, images, and interactivity as a way to demonstrate their learning. User-generated content tools now offer students a platform for creative expression and an authentic audience that enables them to experience the social power of information and communication quite directly.

However, many of the typical approaches to online and distance learning are clearly inferior to the rich and dynamic learning experience of the seminar room. Few asynchronous activities promote the kind of collaborative engagements that activate intellectual curiosity. What could be more boring than activities such as reading from a repository of articles posted online, participating in a threaded discussion, viewing prerecorded webcasts or online lectures, engaging in web quest activities that involve reviewing preselected information sources to find information, and taking online quizzes? Fortunately, synchronous activities, including online text or video chat, whole-class or small-group conferencing, multimedia composition, and the use of social media can help support learners by providing a more active, collaborative, and emotionally supportive learning environment.

The educational technology research literature is rife with studies that find no differences between learning experiences that use technology and those that do not. A meta-analysis review of more than 1,000 studies of online and distance learning revealed that learning effects are larger when online and face-to-face instruction are blended (Means, Toyama, Murphy, Bakia, & Jones, 2010). Neither the use of online quizzes nor the use of video enhanced learning. Instead, it's all about control and metacognition, because "online learning can be enhanced by giving learners control of their interactions with media and prompting learner reflection. Studies indicate that manipulations triggering learner activity or learner reflection and self-monitoring of understanding are effective when students pursue online learning as individuals" (Means et al., 2010, p. xvi).

As a communication scholar with interests in education, I am fascinated by how to promote the transfer of learning so that the online course experience builds knowledge and skills applicable to the world outside the classroom. For this reason, the *open network learning environment* has been most important to me because the Internet and open-source digital tools, such as Twitter, YouTube, and WordPress, can be used as a means to escape the walled garden of the learning management system (LMS). "Open network design enables students to build an authentic network learning community through context-rich social interaction rather than focusing on content only" (Tu, 2014, p. 147). Whereas LMSs such as Blackboard, Canvas, and Sakai enable secure access to informational content, the skills learned in using an LMS for learning have limited transferability to the world outside the classroom. In contrast, the skills learned in using open-source tools, creating user-generated content and participating in knowledge communities using social media tools for learning, are immediately generalizable to workplace and community.

Connecting as Learning

Today, learning may be more important than knowing. Indeed, the legacy of early 20th-century Russian education scholar Lev Vygotsky helps account for how learning happens in the 21st century. Wertsch (1985) described Vygotsky's conceptualization of human activities as complex, socially situated phenomena, where both the individual subject and the social reality exist in systemic context. As digital media and technology become an increasingly ubiquitous part of our cultural environment, with information and opinions at our fingertips, knowledge seems to exist "out there" in the world rather than in the mind of an individual (Siemens, 2005). The concept of connectivism has emerged to argue that relationships among people, ideas, and information are at the heart of learning practice. As Evans (2014) noted, "technologies that facilitate connections between people and information resources should enhance learning because knowledge is a product of these connections rather than simply what is in the head of the learner" (p. 913). Similarly, building on the concept of participatory culture (Jenkins et al., 2006), Ito and her colleagues (2013) have advanced the concept of connected learning as a type of learning that integrates personal interest, peer relationships, and achievement in academic, civic, or career-relevant areas.

There is, at present, a slender research literature on the pedagogical uses of Twitter, so evidence is mixed about its value. Using Twitter for teaching may help students engage in course-related activities by increasing student engagement (Chen, Lambert, & Guidry, 2010; Grosseck & Holotscu, 2008). In one case, a field experiment was used to assess how Twitter may affect undergraduate students' engagement in a health course; researchers found that Twitter users reported more cultural and community engagement and informal discussion of class topics with peers (Junco, Helbergert, & Loken, 2011). Junco, Elavsky, and Heiberger (2012) also found that required Twitter usage by students and staff was associated with student engagement. Evans (2014) examined 252 undergraduate students in a business and management class who were encouraged to use Twitter for communicating with the instructor and peers. A factor analysis revealed that students who used Twitter more frequently had higher levels of engagement in university-associated activities and sharing information. When participants of a week-long professional development conference were asked to tweet about their experiences during the week, content analysis revealed that many partici-pants continued discussions on the topics of the conference. Postconference evaluations showed that participants valued the ability to engage in immedi-ate communication and share information with others (Costa, Benham, Reinhardt, & Sillaots, 2008). However, some research indicates that the novelty of Twitter may interfere with content mastery: For example, under-graduate students who were asked to use Twitter to communicate and share information appeared to only gradually learn to incorporate academic, course-related content (Ebner, Lienhardt, Rohs, & Meyer, 2010).

When using Twitter and other social networks for learning, issues of time, technological awareness, and structures of authority all contribute to

challenge the shift from a focus on platforms focused on delivery to platforms and infrastructures facilitating dialogue, exchange, and collaborative
participation. Researchers have documented how educational institutions
only gradually shift from a closed community with a particular focus on
teaching and learning to a more open community that connected with its
wider networks (Clark, Couldry, MacDonald, & Stephansen, 2014). Of
course, this work can be challenging to teachers and learners alike.

Producing Tweets for Online Learning

For three semesters in 2013 and 2014, I explored Twitter to support my
teaching in online graduate-level courses (12–18 students per course) at the
University of Rhode Island. During this time, and with the permission of my
students, I used a digital tool (www.seen.co) to collect and archive samples
of tweets produced by students at various times during the semester. I would
occasionally display the archive to students and comment on the patterns I
was observing. Although I collected and analyzed course-related tweets for
teaching purposes, the sample is a representative array from more than 200
tweets generated during the course of three semesters. A number of students
commented in their reflective writing that the experience of having to learn
to use Twitter was valuable to them, but others described the experience as
uncomfortable. The tweets reported in this chapter were created by students
who either responded to a specific assignment-related request from the
instructor or composed tweets as a freely chosen activity, and although
tweets are used with the students' permission, I have masked the user names.

In what follows, I focus my attention on *the nature of the learning curve*
in using Twitter. Such a focus supports the generation of reflective thinking
about teaching and learning and potential areas of inquiry for future exploration. As digital media and technologies become essential dimensions of
learning and teaching, educators must consider the learning progression
involved in the effective use of Twitter as an information and communication
tool. In the section that follows, I examine students' ability to use Twitter to
summarize and analyze the course content, including readings and videos. I
explore how Twitter supports peer-to-peer interaction and collaboration and
how students may develop professional relationships with individuals
beyond the classroom. I look at students' ability to share information content
and other relevant resources and engage in self-promotion and self-advocacy
and reflect on my own use of Twitter to call attention to exemplary work.

Summarize and Analyze New Ideas

Twitter's mode of expression mandates that messages be crisp and concise. When first beginning to use Twitter, students were challenged by the forced concision of 140 characters. For this reason, learners were encouraged to practice concision by summarizing ideas from the course reading, viewing, and listening activities. Sometimes I would model the elements of a concise Tweet, as in this summary of a key idea from the work of David Lankes, whom we were reading that semester:

> Lankes: The shift against hierarchy on the Internet is perfectly matched to teens' own internal shift against authority #LSC531

Some learners struggled to move from beginner to intermediate stages in one semester, finding it difficult to express themselves concisely and coherently. For example:

> #LSC530 YALSA says-literacy=more than cognitive ability to read&write but=social act that involves basic modes of participating in world!!!

It's possible that this tweet was comprehensible to the members of the class who read it, but it does not represent effective skills of summarizing or sharing ideas. Students' tweets sometimes demonstrated their capacity to use concision in ways that, while capturing key ideas, in some cases compromised readability. For example:

> Lankes said library should be a platform for innovation and new librarianship is about being active. Key words - platform & active! #LSC531

> Halo effects of authority/reliability: authority in one area presumed authority in others, automatically. Assessing info is harder! #LSC531

In general, over time, students were able to capture the ideas of authors by summarizing effectively. Some students even took a special thrill in finding the author's Twitter handle and including it in the summary:

> For young people, the Internet may provide the resources for negotiating the balance between conformity & rebellion #LSC530 @NAME10

A few students were even able to use tweets to share critical reflection on the reading, building argument chains. This student acknowledged a key argument presented in the reading while offering a practical critique:

> I have conflicted feelings about the YALSA piece. I'm drawn to a welcoming community where teens can be themselves #LSC530 #media #children

BUT the realistic side of me keeps screaming BUDGET! MONEY! I think you could implement some of these practices & build later #LSC530 #media

Engage in Peer Social Interaction

Twitter is a powerful tool for peer-to-peer relationship development. After students had mastered the practice of summarizing with concision, I encouraged them to respond to the ideas of their peers. Although some used Twitter for interpersonal dialogue, many chose to make some forms of their social interaction more accessible to the whole class by including the course hashtag (#LSC531, #LSC530) that enabled students to easily monitor whole-class activity. Some tweets were requests for information or clarification:

@reneehobbs and #LSC531, I'm in search of some data-driven studies re: why school libraries matter, to share with admins. Suggestions?

#LSC530 Need help! Does anyone know how to access LibGuide thru URI so I can start my final project.

Another common use of Twitter was in the provision of warm feedback to support the learning process:

@NAME1 @NAME2 @nytimesbooks Nice job ladies! Beautiful layout & I liked how you included how the editor connects via SM. #LSC530

@NAME3 @NAME4 I hadn't ever thought to read the @nytimesbooks for children/YA reviews. Thanks, and nice @padlet! #LSC530

These forms of emotional support were common; this reflects an appreciation of peer-to-peer learning as a form of cultivating lifelong learning through the development of professional relationships.

Develop Professional Relationships Beyond the Classroom

Twitter enables people to use the power of personal learning networks for learning because users follow as people who offer perspectives and insights that are perceived to have value. As they developed confidence in using Twitter, I required students to develop a network of at least 100 people to follow. However, most students did not engage with or actively reach out to unknown professionals on Twitter, preferring instead to interact with peers enrolled in the course. But a few students discovered that they could develop relationships with professionals by using their names in their tweets. For example, some students tweeted thanks to guest speakers, as in this case:

Watching class now, @NAME5 thank you for joining our #LSC531 and talking about the NARA project you and many others are hard at work on!!

I may not be able to fully understand whether and how students developed professional relationships beyond the classroom because of the limitations of my research method, which uses only posts tagged with the course numbers (#LSC530, #LSC531). Students may have engaged in dialogue with other Twitter users without marking these interactions as class-related through hashtagging.

For communication, media, library, and information studies students, in particular, a deeper understanding of hashtags was an essential dimension of the course experience. Hashtags have been formally part of Twitter since 2009, serving as a self-organizing taxonomy that enables content creators to label their tweets and posts so that others can easily search for related posts. Thus, #medialiteracy will retrieve tweets about media literacy and #publiclibrary will find the tweets about public libraries.

Hashtags also link people together and serve as a sort of unmoderated discussion forum. For example, a group of educators uses the hashtag #digicit to hold regular synchronous conversations on digital citizenship at a designated time period. Although hashtags are neither registered nor controlled by any one user or group of users and do not contain any set definitions, hashtags support engagement by groups of people who participate in conferences or special events. Through hashtags, people can connect with particular special interest groups or members of a discourse community.

In my class, some students used hashtags skillfully for communicating with professional networks by recognizing and using terms like #yalsa (the Young Adult Library Services Association) to signal their interest in addressing adolescent or young adult library service issues. But for other students, the process of learning to recognize the discourse communities associated with different hashtags was accomplished through a process of trial and error. For example, without prompting, one student began using the hashtag #kidlit in his tweets. He responded to a class discussion exploring dystopic themes in young adult literature and current films. He wrote:

#kidlit has always prepared kids for the future. Why is the current strategy to rattle them with the threat of dystopia? #lsc530 #hungergames

As a new entrant into the discourse community, this student was unaware that the #kidlit hashtag is used by publishers, authors, librarians, and readers of contemporary children's literature, with special attention to children ages 5 to 11. Had the student used the #yalit hashtag, he would have found a more targeted audience of young adult librarians and authors, including many with specific interest and expertise in dystopic themes in young adult literature targeting adolescents. But the student was less likely to find these users with the #kidlit hashtag. Fortunately, the use of the #hungergames hashtag did connect the student to the fan community of the Suzanne Collins adventure series and the film adaptations that were released beginning in 2012. Over time, and with sufficient opportunity for exploration, students did gain familiarity with the nuances of hashtags associated with the most relevant interest-driven communities on Twitter.

Share Informational Content or Other Relevant Resources

Twitter is a powerful tool for information discovery, access, and sharing. Students experienced a sense of wonder about the variety of resources they discovered using Twitter, sharing resources that were valuable and relevant and learning to construct meaningful tweets that had value to others. They composed tweets intentionally to share and interpret the resources that they found. For example:

> Here's the @CNN article on HOMAGO: Making the library cool for teens http://us.cnn.com/2014/06/02/living/library-learning-labs-connected-learning/index.html?sr=sharebar_twitter ... #LSC531

> Mass media criticism in the 1800s! "The content of popular novels was wrecking the values of the masses" arenastage.org/shows-tickets/... #LSC530

In the former example, the student shared a mass media resource that addressed a topic discussed in the class. The latter reflects an explicit connection between the student's experience of a local theater production as it relates to some course themes. Another student shared a YouTube video link:

> It's amazing how creative people can be when adapting their favorite series for mass production #lsc530 https://t.co/uoElfm8Hjz

This student discovered this video by exploring the work of the youth adult author Marissa Meyer and tweeted a link to a YouTube video series, *Cinder,* based on Meyer's young adult series of the same name. However, this student did not interrogate the authorship of the video. If he had, he would have discovered that it was made by a 16-year-old girl.

However, mastering the norms of Twitter for information sharing involves a learning process in itself. Analysis of less effective tweets helps me explore and understand dimensions of that learning progression. For example, some students initially seemed confused by the distinction between the

#hashtag and the @username. For example, one student summarized an idea from a course reading (in this case, the work of the teacher-librarian author Amy Pattee) by composing this tweet:

> #Pattee collection dev. represents the mission and goals of the institution, take a look at ALA's guidelines:ala.org/tools/atoz/Col... #LSC531

Here the student used a hashtag for the author's name instead of using Amy Pattee's username, @bapattee. The student also linked to a page of resources that was so broad and general as to be of very little value to someone looking for specific quality resources. This student was challenged to identify a relevant resource associated with this topic. In another example, a student clearly intended to share information with her peers but did not include the hyperlink to the readily available online resource:

> #LSC530 Voya Vol. 36, #6, Feb 2014 is all about Teens & Technology

Errors of this type may simply be normative for users, reflecting the speed and ease of using Twitter, or it may suggest that learning progression could be aided by explicit instruction.

Engage in Promotion and Advocacy

Twitter offers a way for people to exercise their voice as community-engaged professionals and to advocate for their own needs and goals. As one who bridges the worlds of theory and praxis, I modeled the use of Twitter for self-promotion during the semester as I shared news about upcoming publications, presentations, workshops, and curricular innovations. As part of course assignments, I also asked students to tweet links to their completed assignments, given that all work was published online. This served to encourage peers to read and respond to each other's work, and it also cultivated a sense of public authorship among students who may have perceived that only the instructor was reading their writing.

Learners need confidence and high levels of self-esteem in order to speak with a public voice. But some learners did not feel comfortable or confident about expressing ideas in a public forum. In fact, not all students even complied with my request to share URLs to their work using Twitter. For some, the idea of sharing their work may be intimidating. For those who participated in sharing their work, these tweets generally took a form such as this:

> Informing and the Digital World http://t.co/Jsn3VJlo4u via @NAME6 final creative project for #LSC530

In general, students provided warm feedback to their peers, suggesting the development of a supportive and respectful community. Some students even used Twitter actively to solicit peer feedback for their work:

> Anyone from #LSC530 watched my ignite presentation on #yabookreviews please let me know what you think via tweet. Feedback is appreciated

Because of its public nature, Twitter may support the development of a professional identity that enhances student confidence and supports creative expression. As evidence to support this, I observed that occasionally students would share personal projects they developed outside of a class assignment if these were perceived to be relevant to the interests of the class members:

> Need ideas for children or teen programming? Take a look at my boards on Pinterest at:pinterest.com/NAME7/. #LSC530 #children #library

> My YouTube multimedia playlist to promote literature to students: youtube.com/playlist?list=... #LSC530

In the first case, the student shared a well-curated collection of mostly arts and crafts projects that may be suitable for home or library activities. In the second, the student curated YouTube videos, some that were personally collected and others that were used in the graduate course. In doing so, the student demonstrated the effective use of the YouTube Playlist for curation.

Sometimes students used Twitter for self-advocacy to address their needs as learners in the course, as in the case of a student who used Twitter to gauge student opinion about the merits of adjusting the deadlines for a particular assignment. In this case, the student created a digital poll and embedded it in her tweet to gather data on student opinion:

> Are you in favor of flip-flopping the next two assignments for #LSC530? polleverywhere.com/multiple_choic... via@polleverywhere

Of course I was thrilled to see this example of using Twitter to inspire advocacy. So I changed the deadline as students requested, with my general message of "Ask and you shall receive."

Provide Attention as a Reward

Twitter offers a way for instructors to showcase and celebrate exemplary work and to offer public feedback in a way that can inspire and motivate learners. As an instructor, I struggled with how to provide feedback to advance student engagement in the fully online course. Students received private feedback on their performance in course assignments through e-mail.

But how could I offer feedback in a more public way to identify the highest caliber of student performance and help members of the class to recognize excellence? I began by signaling my appreciation of student tweets by using the *favorite* button, which indicates to students that I had viewed the tweet. Over time, I began to use Twitter to showcase exemplary examples of student sharing and participation:

"Collection development in YA is a political act." #LSC531 #libraries @janery8 VIDEO: flipgrid.com/#4924d2b0

Here I tweeted a student video comment from an online discussion about the controversies associated with collection development in public and school libraries, sharing a video link to a classroom discussion where we used Flipgrid, which is a digital discussion tool allowing participants to contribute 90-second video comments in an online discussion. Some examples of my feedback to students include these tweets:

The security of books & the risks of online exploration: a fascinating essay by @NAME 8 bit.ly/1ece3pz #LSC530 #mystudentsareamazing

Children's literature has a long tradition of being subversive says @NAME9 #LSC530 #mystudentsareamazing VIDEO: http://bit.ly/1AntOr6

Celebration of exemplary student work serves to set the bar for all students and offers models of writing and creative work that may inspire others. At various points in the semester, I described and shared a link to a piece of student writing and video production, using the hashtag #mystudentsareamazing, which came to be my approach to using attention as a reward.

The Challenge of Leaping from Classroom to Community

In 1965, describing the process of curriculum design, Jerome Bruner (1965) wrote, "As for stimulating self-consciousness about thinking, we feel that the best approach is through stimulating the art of getting and using information—what is involved in going beyond the information given and what makes it possible to take such leaps" (p. 21). This idea continues to be a key dimension of my work in supporting learners of all ages. By exploring how Twitter can be used to stimulate the art of getting and using information to go beyond the information given in the context of a fully online graduate-level course, I discovered that graduate students appreciated the chance to deeply engage with Twitter and learn how to use it. As they summarized and analyzed the course content and engaged in peer interaction and collaboration, they deepened both their understanding of the material and their sense of community with their peers. Considering the challenges of online learning, with its often faceless and joyless characteristics, Twitter was useful in advancing authentic learning by creating a warm and supportive learning environment for busy adult learners who were juggling school, careers, and family lives.

But in reflecting on three semesters of exploring the use of Twitter as a pedagogical tool, there is plenty of room for future development. In the following sections, I explore some of the possibilities.

Twitter in Formal and Informal Learning

Twitter is a powerful tool for informal, self-directed learning, and it has considerable value in online formal learning contexts. Students were able to use Twitter as required to fulfill course expectations. But under some circumstances, for some learners, the authority structures of school may combine with the public nature of Twitter to discourage students from engaging in playful or genuinely dialogic experimentation with ideas. In an open network learning environment, students concerned about issues of reputation and privacy may feel reticent to enact the same kind of healthy risk taking that happens around the university seminar table (a more private space, in which students explore ideas they may not have yet fully formulated). Perhaps because of the public nature of the online space, norms of politeness were present among my students, who generally offered little feedback that was not cheerful, positive, and supportive. Generally speaking this is a good thing. But although it is a little uncomfortable to admit this, the lack of interpersonal conflict was disappointing to me because of my general belief that the frisson of intellectual tension is generally good for provoking people to rethink and question ideas. It's also the case that the students— graduate students enrolled in either a master's program in communication studies or library and information studies—were a rather homogeneous group of people who perhaps share a common worldview and have similar life experiences. This may naturally have limited differences of opinion. It's important to note that arguing in a seminar room is different from arguing in public, and further reflection will be necessary to understand how Twitter shapes or limits academic argument in the context of online learning.

Twitter for Civic Engagement

Although Twitter is widely recognized as a tool for mobilizing public action and advocacy, I did not consciously model the use of Twitter for civic engagement as part of the course. I used Twitter as a pedagogical tool to support mastery of course content and to build a deeper sense of peer-to-peer connectedness. In future work, I will explore how to introduce students to the power of Twitter for civic activism. Although we accessed and analyzed a considerable range of issues related to the professional library and information studies community, I did not require students to participate in advocacy related to controversial topics of public interest with relevance to both the course and the larger society. This is not because there were no such controversies. During the time in which I was teaching with Twitter, concerns about surveillance and monitoring of school Internet, discussions of net neutrality, gender stereotypes in children's media, increased funding cuts for school libraries, parental support for children under age 14 using Facebook and other social media, cyberbullying and online stalking, the rise of open-access publishing models and their impact on the publishing industry, and the popularity of children's literature written (or ghostwritten) by celebrities were some of the many controversies that were part of the public sphere. As an active Twitter user, I contributed to dialogue about these (and other) issues, but I did not explicitly link this practice to the graduate courses I was teaching. Perhaps these issues did not enter the course's Twittersphere in part because I did not provide the explicit scaffolding to support it.

The practice of scaffolding involves more experienced users modeling behaviors to introduce them to less-experienced users. In scaffolding, the mentor enables "the performance of a more complicated act than would otherwise be possible" until the learner is able to accomplish the activity independently (Pea, 2004, p. 425). By consciously modeling engagement in larger social and political issues with Twitter, educators may help students activate practices of authentic civic engagement. This is a key concept of digital and media literacy, which includes the practice of *action* as a key part of the literacy spiral that also includes access, analysis, creation, and reflection (Hobbs, 2010).

I am also interested in how Twitter may be used to increase reflection and metacognition among learners. Pedagogies that support the reflexive process of critical inquiry about the use of Twitter will be important to advance this goal. For example, students could be invited to analyze the characteristics of the individuals they have selected to follow. Because users are responsible for selecting whom they follow, they themselves may inadvertently create filter bubbles, those narrow niche communities that provide little opportunity to encounter the diverse and sometimes challenging perspectives of others (Pariser, 2011). This is a topic of special relevance to future librarians and information professionals, who have sometimes been accused of insularity and navel gazing that is not sufficiently connected to larger communities and networks (Dupuis, 2010; Kim, 2010).

In exploring the public nature of Twitter as a resource to advance life-long learning, it will be important for educators to understand how Twitter may *help people see themselves as lifelong learners, social actors, and narrators of their individual lives.* Although beginners, my students were participating in civic culture, even if rather narrowly defined within the context of our small classroom. Civic culture is an analytic construct that seeks to identify the possibilities of people acting in the role of citizens, where values, affinity, knowledge, practices, identities, and discussion all play a part (Dahlgren, 2005). It will be important to help students recognize the usefulness of small acts of expression and communication that contribute to building a civic culture. Other researchers have recognized the gradual way in which this process occurs in the context of online learning with Twitter. In a qualitative study of the use of digital storytelling for civic engagement, British students who were encouraged to use Twitter first conceptualized its value solely in terms of marketing and promotion: "Students' apparent inability to comprehend the Twitter event in terms other than promotion makes sense given an absence of shared spaces for dialogue beyond the classroom and curriculum: students lacked familiarity with the idea that such dialogue might be encouraged" (Couldry et al., 2014, p. 624). Although the classroom space includes opportunity for resistance, learners can experiment with "a digitally enabled circuit of civic culture" (Couldry et al., 2014, p. 628), even when extending beyond the purely local can be challenging. According to the adage, the personal is the political. As a powerful digital tool for cementing social relationships built on respect and trust, the use of Twitter can help create and maintain social ties within and beyond the classroom community that enable meaningful learning.

References

Berr, J. (2014, July 1). Student loan interest rates rise, worrying some experts. *CBS News Market Watch.* Retrieved October 1, 2014 from http://www.cbsnews.com/news/student-loan-interest-rates-rise-worrying-some-experts/

Bruner, J. S. (1965). *Man: A course of study* (Occasional paper no. 3). Washington, DC: National Science Foundation. Retrieved from ERIC database. (ED178390)

Buchem, I. (2012). Serendipitous learning: Recognizing and fostering the potential of microblogging. *Formare, 74*(11), 7–16.

Chen, P. D., Lambert, A. D., & Guidry, K. R. (2010). Engaging online learners: The impact of web-based technology on college student engagement. *Computers & Education, 54*, 1222–1232.

Clark, W., Couldry, N., MacDonald, R., & Stephansen, H. (2014). Digital platforms and narrative exchange: Hidden constraints, emerging agency. *New Media and Society* [online]. doi:10.1177/1461444813518579

Costa, C., Benham, G., Reinhardt, W., & Sillaots, M. (2008). *Microblogging in technology enhanced learning: A use-case inspection of PPE summer school 2008.* Paper presented at the European Conference on Technology Enhanced Learning (ECTEL) 2008. Retrieved January 25 2014 from http://know-center.tugraz.at/download_extern/papers/2008_ccosta_microblogging.pdf

Couldry, N., Stephansen, H., Fotopoulou, A., MacDonald, R., Clark, W., & Dickens, L. (2014). Digital citizenship? Narrative exchange and the changing terms of civic culture. *Citizenship Studies, 18*(6–7), 615–629.

Cuban, L. (1993). Computers meet classrooms: Classrooms win. *Teachers College Record, 95*(2), 185–210.

Dahlgren, P. (2005). The internet, public spheres, and political communication: Dispersion and deliberation. *Political Communication, 22,* 147–162.

Dupuis, J. (2010, May 12). The inherent insularity of library culture? *ScienceBlogs.* Retrieved January 5, 2015 from http://scienceblogs.com/confessions/2010/05/12/the-inherent-insularity-of-lib/

Ebner, M., Lienhardt, C., Rohs, M., & Meyer, I. (2010). Microblogs in higher education: A chance to facilitate informal and process-oriented learning? *Computers & Education, 55,* 92–100.

Evans, C. (2014). Twitter for teaching: Can social media be used to enhance the process of learning? *British Journal of Educational Technology, 45*(5), 902–915.

Grosseck, G., & Holotescu, C. (2008). *Can we use Twitter for educational purposes?* Paper presented at the e-Learning and Software for Education conference, Bucharest, Romania. Retrieved April 11, 2015 from http://bit.ly/1EFzCBc

Hobbs, R. (2010). *Digital and media literacy: A plan of action.* Washington, DC: John S. and James L. Knight Foundation and Aspen Institute.

Ito, M., Gutiérrez, K., Livingstone, S., Penuel, B., Rhodes, J., Salen, K., . . . Watkins, S. C. (2013). *Connected learning: An agenda for research and design.* Irvine, CA: Digital Media and Learning Research Hub.

Jenkins, H., Purushotma, R., Clinton, K., Weigel, M., & Robinson, A. (2006) *Confronting the challenges of participatory culture: Media education for the 21st century.* Chicago, IL: The MacArthur Foundation.

Junco, R., Elavsky, C. M., & Heiberger, G. (2012). Putting the test: Assessing outcomes for student collaboration, engagement, success. *British Journal of Educational Technology, 44*(2), 273–287.

Junco, R., Helbergert, G., & Loken, E. (2011). The effect of Twitter on college student engagement and grades. *Journal of Computer Assisted Learning, 27,* 119–132.

Kim, J. (2010, April 27). Some takeaways from "this book is overdue." *Technology and Learning Inside Higher Ed.* Retrieved January 4, 2015 from https://www.insidehighered.com/blogs/technology_and_learning/some_takeaways_from_this_book_is_overdue

Korrki, P. (2014, May 24). The ripple effects of rising student dept. *The New York Times.* Retrieved August 1, 2014 from http://www.nytimes.com/2014/05/25/business/the-ripple-effects-of-rising-student-debt.html?_r=0

Means, B., Toyama, Y., Murphy, R., Bakia, M., & Jones, K. (2010). *Evaluation of evidence-based practices in online learning: A meta-analysis and review of online learning studies.* Washington, DC: U.S. Department of Education, Office of Planning, Evaluation, and Policy Development.

Mollett, A., Moran, D., & Dunleavy, P. (2011). Using Twitter in university research, teaching and impact activities. *London School of Economics Public Policy Group.* Retrieved September 28, 2014 from http://blogs.lse.ac.uk/impactofsocialsciences/files/2011/11/Published-Twitter_Guide_Sept_2011.pdf

Pariser, E. (2011). *The filter bubble.* London, UK: Penguin.

Pea, R. D. (2004). The social and technological dimensions of scaffolding and related theoretical concepts for learning, education and human activity. *The Journal of the Learning Sciences, 13*(3), 423–451.

Siemens, G. (2005). Connectivism: A learning theory for the digital age. *International Journal for Instructional Technology and Distance Learning, 2*(1), 3–10.

Thompson, C. (2007, June 26). Clive Thompson on how Twitter creates a social sixth sense. *Wired Magazine, 15*(7) [online]. Retrieved January 31, 2015 from www.wired.com/techbiz/media/magazine/15-07/st_thompson

Tu, C. H. (2014). *Strategies for building a Web 2.0 learning environment.* Santa Barbara, CA: Libraries Unlimited.

Wertsch, J. (1985). *Vygotsky and the social formation of mind.* Cambridge, MA: Harvard University Press.

Engaging Adolescents in Narrative Research and Interventions on Cyberbullying

Heidi Vandebosch, Philippe C. G. Adam, Kath Albury,
Sara Bastiaensens, John de Wit, Stephanie Hemelryk Donald,
Kathleen Van Royen, and Anne Vermeulen

"A few years ago . . . . It was the whole class against one person . . . . And then we made a website . . . . And that girl . . . . She talked strangely. She was annoying. And her knees looked orange . . . . And we named her 'the carrot' . . . . And then we made a website . . . with pictures of her . . . 'the walking carrot' . . . (laughs)" (girl, 17 years old). (Vandebosch & Van Cleemput, 2008a, p. 22)

Several large-scale studies in the United States (Madden et al., 2013), Europe (Hasebrink, Livingstone, & Haddon, 2008), and Australia (Green, Brady, Ólafsson, Hartley, & Lumby, 2011) show that information and communication technologies (ICTs) provide adolescents with a range of opportunities for learning and for increasing creativity, social contacts, and civic engagement. These studies, however, illustrate that computers, mobile phones, and the Internet also present various types of risks. In this chapter, we focus on cyberbullying, a specific type of aggression occurring in online interaction, which can involve adolescents as victim, perpetrator, or bystander. Cyberbullying has been associated with negative health outcomes. Studies have shown, for instance, that cyberbullying victims experience increased levels of emotional distress (e.g., Mishna, Khoury-Kassabri, Gadalla, & Daciuk, 2012; Şahin, 2012; Šléglová & Černá, 2011), depression (e.g., Kowalski & Fedina, 2011; Machmutow, Perren, Sticca, & Alsaker, 2012; Schneider, O'Donnell, Stueve, & Coulter, 2012), and anxiety (e.g., Dempsey, Sulkowski, Nichols, & Storch, 2009; Juvonen & Gross, 2008; Kowalski & Limber, 2013; Pure & Metzger, 2012) and are more inclined to engage in self-harming behavior (Price & Dalgleish, 2010; Schneider et al., 2012) and commit suicide (Schneider et al., 2012).

As we show later, most studies on cyberbullying are quantitative in nature, emanating from adult researchers' viewpoints. In this chapter, we argue that qualitative approaches, in particular those using narratives and involving adolescents' perspectives, should complement quantitative insights to increase understanding of cyberbullying and inspire effective interventions to address this problem. The interview excerpt at the beginning of this chapter, for instance, might provide input for a realistic movie scenario, aimed at raising awareness among adolescents. It also indicates what type of online content might be considered cyberbullying and in this way informs modes of detection of cyberbullying. Finally, this cyberbullying story, especially when told from the victim's perspective, might help young people with similar experiences cope with their situation.

Overview of Cyberbullying Research

Research on cyberbullying started around 2004 (Smith & Steffgen, 2013). Building mainly on research on traditional bullying and the criteria associated with this type of aggression, such as intent to harm, repetition, and power imbalance (Olweus, 1993), cyberbullying was defined as "An aggressive, intentional act carried out by a group or individual, using electronic forms of contact, repeatedly and over time against a victim who cannot easily defend him or herself" (Smith et al., 2008, p. 376). More concretely, cyberbullying included a wide range of behaviors such as sending threatening messages, distributing embarrassing pictures, creating false profiles, and excluding others online through computer or mobile phone applications (Vandebosch & Van Cleemput, 2009).

Typically, early empirical studies focused on measuring the prevalence of cyberbullying, establishing profiles of bullies and victims, and relating cyberbullying victimization to possible mental health correlates through large-scale, cross-sectional studies (for an overview, see Kowalski, Giumetti, Schroeder, & Lattanner, 2014). Some standardized questionnaires used direct measurements, asking youngsters whether they had (or had been) cyberbullied, often even without defining the term (Heirman & Walrave, 2008; Pyzalski & Wojtasik, 2010; Sourander et al., 2010; Vandebosch, Van Cleemput, Mortelmans, & Walrave, 2006). Others used indirect measures, providing adolescents with a list of concrete actions (e.g., sending threatening messages, hacking into someone's account) supposed to represent cyberbullying (Gámez-Guadix, Orue, Smith, & Calvete, 2013; Menesini, Nocentini, & Calussi, 2011; Salmivalli, Sainio, & Hodges, 2013; Sticca, Ruggieri, Alsaker, & Perren, 2013; Vandebosch & Van Cleemput, 2009). Different types of measurements, in different types of surveys, aimed at different age groups, resulted in highly varying prevalence rates, inconsistencies in the profiles of cyberbullies and victims, and variations across studies with regard to the strength of the correlations between cyberbullying involvement and mental health indicators (Kowalski et al., 2014; Tokunaga, 2010). The ambiguity in the operationalization of cyberbullying thus proved to be problematic in comparing data both internationally and over time.

In addition, a debate arose regarding whether scholars' definition of cyberbullying reflected adolescents' perspectives on this phenomenon or the lived experience of individuals involved, whether cyberbullying victims, perpetrators, bystanders, parents, school personnel, or others. Attempts to address these concerns were undertaken by a dozen qualitative studies that mainly relied on semistructured focus groups or in-depth interviews. Vandebosch and Van Cleemput (2008b), for instance, conducted 53 focus groups among 12- to 18-year-olds in Flanders, Belgium, and concluded that some Internet or mobile phone practices typically included in questionnaires to measure cyberbullying indirectly did not necessarily constitute true cyberbullying for young people, whose interpretation of cyberbullying is highly context dependent. The importance given to context by young people aligns with Olweus's (1993) definition of traditional bullying. To be classified as cyberbullying, the Internet or mobile phone practices should be intended by the sender to hurt, be part of a repetitive pattern of negative offline or online actions, and be performed in a relationship characterized by a power imbalance (based on real-life power criteria such as physical strength, age, social status, and/or on ICT-related criteria such as technological know-how and anonymity). These criteria helped distinguish cyberbullying from cyberteasing (not intended to hurt, not necessarily repetitive, and performed in an equal-power relationship) and cyberarguing (intended to hurt, not necessarily repetitive, and performed in an equal-power relationship; Vandebosch & Van Cleemput, 2008b).

Marwick and boyd (2011, 2014) came to similar conclusions on the basis of semistructured interviews with 166 teenagers in the United States. They stated that many practices commonly considered cyberbullying by adults were in fact described by girls as "drama," a notion that refers to relational conflicts between equals (a reciprocity that also made it distinct from bullying) that were performed online (although they could have started offline), in front of and magnified by networked audiences. Drama referred to practices ranging from gossiping, flirting, arguing, and joking to more serious issues of jealousy, ostracizing, and name calling. Whereas boys were often the subject of drama and also watched it from a sideline, they mostly distanced themselves from these feminine practices. Boys, on the other hand, more often engaged in what they considered masculine practices, such as punking or pranking (performing mischievous acts). According to Marwick and boyd (2011, 2014), the dynamics of drama, punking, or pranking were thus different from those described in bullying narratives. Marwick and boyd (2014) urged that research and interventions take these distinctions into account: "To support youth as they navigate aggression and conflict in a networked society, adults must begin by understanding teenage realities from teenage perspectives" (p. 16).

The realities of what it means to be the victim, perpetrator, or bystander of cyberbullying and related forms of aggression have been partially explored by Šléglová and Černá (2011), Varjas, Talley, Meyers, Parris, and Cutts (2010), and DeSmet et al. (2013). Šléglová and Černá interviewed 15 adolescents (ages 14–18) who self-identified as victims of what the researchers considered cyberbullying via social network sites. These interviews conducted via ICQ or Skype chat revealed that cyberbullying varied in terms of duration, from several days to more than a year, and in the extent to which it was associated with offline bullying. Victims experienced different degrees of psychological impact, used several types of coping strategies, and varied in the extent to which they adapted their online behavior afterward. Research on possible motives of cyberbullies (Varjas et al., 2010), based on in-depth interviews with 20 high school students (ages 15–20), showed that cyberbullying was mostly perceived as resulting from internal states. Cyberbullying was regarded as a means to redirect one's feelings, take revenge, make people feel better about themselves, alleviate boredom, seek protection or approval, try out a new persona, deal with jealousy, and so on. External motivations, such as the nonconfrontational nature of cyberbullying or the otherness of the victim, were perceived as less important. Finally, a focus group study with adolescents investigating the role of bystanders of cyberbullying (DeSmet et al., 2013), revealed that bystanders or witnesses of cyberbullying could behave in different ways (defending the victim, supporting the bully, doing nothing), both offline and online. Whether and how they reacted seemed not only to be determined by personal characteristics (e.g., the level of empathy) but also by contextual elements. Adolescents mentioned, for instance, that they would act differently according to their relationship (e.g., friends versus acquaintances) with the bully or victim.

These qualitative studies in turn inspired new quantitative studies, which tested definitional issues using vignettes in standardized questionnaires (Menesini et al., 2012) or mapped the most important factors associated with being a victim (Li & Fung, 2012), a perpetrator (Pabian & Vandebosch, 2014), or a bystander (Bastiaensens et al., 2014; DeSmet et al., 2013). Although quantitative studies have provided useful data for evidence-based cyberbullying interventions (e.g., Del Rey, Casas, & Ortega, 2012; Doyle, 2011; Palladino, Nocentini, & Menesini, 2012; Schultze-Krumbholz, Wölfer, Jäkel, Zagorscak, & Scheithauer, 2012), we argue that more qualitative studies, especially narrative studies focusing on youngsters' stories of cyberbullying, have the potential to more directly inspire novel interventions on cyberbullying. Detailed and personal accounts of cyberbullying experiences can provide a unique basis for persuasive communication aimed at addressing and reducing bullying among adolescents.

Collecting Youngsters' Cyberbullying Stories

Narrative methods start from the observation that people often attach meaning to their experiences through stories (Atkinson, 1998) and that storytelling is an important aspect of human communication. Narrative is defined as "a story that tells a sequence of events that is significant for the narrator or her or his audience" (Moen, 2006, p. 60) or as "a representation of connected events and characters that has an identifiable structure, is bounded in space and time, and contains implicit or explicit messages about the topic being addressed" (Kreuter et al., 2007, p. 222). Researchers using narrative methods try to reveal individuals' stories about their subjective experience, with reference to larger sociocultural dynamics and discourses (Parker, 2003; Sutherland, Breen, & Lewis, 2013). This is often done through a collaboration between the researcher and the research subjects (Moen, 2006).

A number of data collection methods can be used in narrative research (Connelly & Clandinin, 1990), with interview-based narrative research the most common (Riessman, 2006). Other approaches involve the use of participatory narrative research, such as digital storytelling (Willox, Harper, & Edge, 2013), or the observation and analysis of naturalistic data, such as narratives that develop on the Internet (Robinson, 2001).

Interview-based narrative research can take the form of *narrative interviewing,* which refers to unstructured interviewing that can either involve the narration of an event in a person's life (narrative inquiry) or a whole life story. What distinguishes the narrative interview from other methods is that it keeps the presentation of the story in the words of the person telling it. Narrative data thus represent a first-person narrative, with the interviewer removed from the text as much as possible (Atkinson, 1998). To stimulate recall or to provide a basis for reflection, materials such as photos or videos can be used in narrative interviews. These stimuli may help to reconstruct past thinking or to construct postactivity narratives or reflections on present and future actions (Jewitt, 2012). In the case of cyberbullying, personal stories can be gathered by interviewing adolescents from a victim, bully, or bystander perspective (or a mix of several, because adolescents may fill more than one of these roles). Stimulus materials might include existing evidence from the adolescents' actual cyberbullying experiences (e.g., print screens), or photos (Walton & Niblett, 2013), online videos (Price et al., 2014), and texts (e.g., newspaper articles) referring to other fictitious or real cases. Games connecting to the bullying theme, such as *Cyberball* (Williams & Jarvis, 2006) or *The Bully* might also be used to elicit personal cyberbullying stories.

The online world, however, can do more than provide textual or audio-visual stimuli to use in offline narrative interviews. ICTs also create possibil-ities to interview youngsters through (a)synchronous communication tools, such as e-mail, fora, instant messaging and chat, VoIP (Voice over Internet Protocol), or in immersive virtual worlds, representing the interviewer and research participant by avatars (Salmons, 2009). Researchers might prefer these interview methods because of several practical advantages, such as cost and time savings or easier handling of data (Mann & Stewart, 2000). Moreo-ver, mediated interview techniques provide specific benefits for young research participants, such as increased confidence for self-conscious youth through visual anonymity (Valkenburg & Peter, 2011) and asynchronous communication tools allowing adolescents to "stop and think before giving a response" (Madell & Muncer, 2007, p. 139).

In *participatory research methods,* respondents are included in the study as research partners and create their story through images, video, audio recordings, or text (Gubrium & Harper, 2013). Bowler, Mattern, Knobel, and Keilty (2014) have used these methods in cyberbullying research involving four teens and five undergraduate students. Participants were probed to tell a story about "mean and cruel behavior online" as they imagined it would be for someone else and to think of interventions that "might alleviate or even intervene in mean and cruel online behavior" at different moments in their story. Spears, Slee, Owens, and Johnson (2009) invited 20 adolescents, 10 teachers, and 6 school counselors to recount their personal stories of cyberbullying. These stories were audio recorded by the participants themselves and then uploaded to a dedicated website, thus contributing to an online storybook.

A last narrative method involves the *collection of naturalistic data.* Individuals may have already expressed their personal stories (e.g., in diaries), without researchers having explicitly elicited them. Rich, spontaneously generated data are also available on the World Wide Web (Robinson, 2001), especially since the emergence of Web 2.0 (Rathi & Given, 2010; Snee, 2008). Possible narrative data sources include blogs (Heilferty, 2011; Hookway, 2008), discussion fora, social networking sites such as Facebook (Hinduja & Patchin, 2008; Jones, Millermaier, Goya-Martinez, & Schuler, 2008), or video sharing sites such as YouTube (Jewitt, 2012; Pace, 2008). In the case of cyberbullying, naturalistic personal narratives can be found on discussion fora of helplines for adolescents (e.g., Kids Helpline). There are also well-known examples of adolescents, especially victims, who have uploaded their personal stories of cyberbullying on YouTube (e.g., the "goodbye video" uploaded by Amanda Todd, who committed suicide after being cyberbullied). Macbeth, Adeyema, Lieberman, and Fry (2013) have retrieved and (partially) analyzed stories on cyberbullying posted on the website "A Thin Line" (http://www.athinline.org/), which is part of an MTV campaign against digital abuse. On this website, users post their stories, and others can indicate whether they think the behavior is socially acceptable. This study discerned several patterns in these stories (such as posting nude pictures of an ex-girlfriend or boyfriend as an act of revenge).

Using Adolescents' Cyberbullying Stories to Inform Interventions

As is clear from the preceding overview, personal stories of adolescents' cyberbullying experiences can be gathered in various ways. In the following sections, we argue that such narratives may be successfully used to develop intervention tools, aiming at general prevention and behavior change and online detection of and appropriate responses to actual cases (i.e., targeted help to victims).

General Prevention and Behavior Change

Several authors (Kreuter et al., 2007; Miller-Day & Hecht, 2013; Stavrositu & Kim, 2014) have argued that, in some instances, narrative health communication might be more effective than nonnarrative (i.e., didactive or statistical) health communication. Whereas the former focuses on description, explanation, and personal experience by representing a sequence of events, characters, and consequences, the latter is about logic, arguments, and statistics. Narrative health communication might be particularly suitable to reach and convince young people, because it renders complex information comprehensible (Hopfer & Clippard, 2011) and helps overcome resistance to more traditional forms of health campaigns. Narrative communication can take different forms: The stories might be fictional or real, told in first or in third person, be more or less interactive (Green & Jenkins, 2014; Hinyard & Kreuter, 2007), and provide greater or lesser amounts of story.

Several theories may explain why narratives might be persuasive. *Transportation theory* (Green & Brock, 2000; Green, Brock, & Kaufman, 2004), for instance, suggests that people can become cognitively and emotionally immersed in a story. This transportation evokes vivid mental imagery, making the story seem more like actual experience; reduces counterarguing, making it more likely for individuals to believe the story propositions; and creates connection with story characters, causing their perspectives to have greater influence on the beliefs of the audience. *Social cognitive theory* (Bandura, 1977) suggests that people learn behaviors by observing role models, such as characters in stories.

Because engagement, realism of the plot, and identification with the characters (Green & Jenkins, 2014) appear to contribute to the persuasiveness of stories, it is important that health communication narratives integrate the experiences of members of the target group. Miller-Day and Hecht (2013) already demonstrated this for an intervention aimed at preventing drug abuse. After collecting adolescent narratives about drug offers, drug refusals, drug use, and the perceptions of the culture of drugs in adolescents' communities, common plot lines and patterned experiences were identified. These reflected several resistance strategies, such as refusing, explaining, avoiding, and leaving. These analyses were used in the message design, in which video stories were produced in cooperation with the target group. Miller-Day and Hecht (2013) believe that involving youth in the process of creating such narratives can serve as a prevention strategy in itself.

In the case of cyberbullying, personal stories can provide bottom-up accounts of the concrete events that adolescents experience, the social context(s) in which these occur, the actions adolescents undertake, and how these influence the outcome of the process for better or worse. Realistic and personally relevant prototype stories in which role models ultimately succeed in behaving correctly (e.g., potential victims taking the right preventive measures or coping effectively with cyberbullying, perpetrators quitting their negative behavior, and bystanders helping the victim), and are rewarded for doing so, may then form the basis for persuasive messages. These stories can, for instance, be used in entertainment-education formats (such as popular teenage series and movies), forms of behavioral journalism (articles in teenage magazines or newspapers), interactive stories on campaign websites, or serious games.

An example of an evidence-based serious game currently being developed to promote positive bystander behavior in cases of cyberbullying is Friendly ATTAC (Vandebosch & Poels, 2012). This game is built on insights from both quantitative and qualitative studies on bystander behavior and its determinants and relies on detailed feedback from the target group (12- to 14-year-olds) in every development and test phase. In an early test phase, for instance, adolescents were asked to provide their comments on several narrative aspects, such as the immersion in the larger story, the realism of the integrated cyberbullying incidents, and the identification with playable and nonplayable characters. It is unclear to what degree other initiatives (such as the interactive cyberbullying scenarios on http://itsuptoyou.nu) rely on evidence of youngsters' cyberbullying experiences.

Although bottom-up stories of cyberbullying might provide realistic and relevant stories for interventions aimed at the larger target group, engaging adolescents in the message production itself might create additional benefits. There are already cyberbullying programs in which the very process of developing and delivering anticyberbullying messages (by using traditional or ICT tools) is considered an effective method to persuade adolescents (Bauman, Cross, & Walker, 2013).

Cyberbullying Detection

Apart from their use as general prevention and behavior change tools, stories of cyberbullying might also provide a basis for early detection of cyberbullying in online environments (such as social networking sites), because they suggest what type of events, in what contexts, to look for. In particular, by connecting personal narratives with concrete online content, information can be gathered on the verbal or visual messages that indicate the onset of certain forms of cyberbullying (e.g., insulting language, embarrassing pictures), their growth (e.g., number of likes by bystanders), or consequences (e.g., a victim expressing distress, adjusting privacy settings; Bowler et al., 2014). This matching of interpretations and manifest characteristics of the situation may also provide further insight into the profiles of youngsters who are more at risk of being a victim or perpetrator of certain types of cyberbullying (based on gender, age, personality, ICT use, online and offline social network, and so forth). By taking into account all these content and context features, automatic detection of possible instances of real cyberbullying (and not of jokes and quarrels) might be possible (Macbeth et al., 2013).

The early detection of cyberbullying allows a timely and adequate response: Perpetrators can be sent a warning message, and victims can be offered support and guidance (Van Royen, Poels, Daelemans, & Vandebosch, 2014). The latter might include references to health professionals or to a tailored selection of stories told by other victims (discussed later).

Responses Aimed at Cyberbullying Victims

Health communication research suggests that being exposed to personal stories from others who have experienced a similar situation might be helpful (Bülow, 2004; Steffen, 1997). Nowadays, many stories are constructed and exchanged online (e.g., http://www.patientslikeme.com). As suggested by Manuvinakurike, Barry, and Bickmore (2013), indexing the huge amount of personal, unstructured health stories from existing repositories of online health information (e.g., blogs, support sites, and personal web pages) can be an important step in providing stories of others that are tailored to one's personal needs and thus are more helpful. Survey-based or automated systems that profile a person and his or her health-related behaviors can in this way be usefully linked to systems containing hand- or automatically coded personal health stories of others.

In the context of cyberbullying, the idea of online tailoring is applied in a traditional way on the website *Pestkoppen Stoppen* ("Stop Bullies"), which is being developed by the Open University in Heerlen in the Netherlands (Jacobs, Völlink, Dehue, & Lechner, 2014). On this website, victims can get tailored advice, mostly in nonnarrative forms, based on their responses to a standardized questionnaire. As suggested earlier, it might be a good idea to add supportive stories of other victims to the professional advice.

Understanding Bullying 2.0 Requires Using
2.0 Narrative Research and Intervention

Cyberbullying remains an important problem and increasingly requires effective interventions. Introducing narrative methods in this mainly quantitative-oriented research field allows a deeper understanding of this complex phenomenon and unveils the meanings assigned to it by adolescents themselves. Researchers may rely on interviews, participatory narrative methods, or the observation of naturalistic data to collect cyberbullying stories. These may then provide the basis for effective general prevention and behavior change and online detection and responses toward victims (see Figure 14.1). The proposed approach thus represents a bottom-up strategy, whereby researchers and intervention developers consult the target group extensively. It also encompasses the use of new technologies to study and address online risks encountered by adolescents.

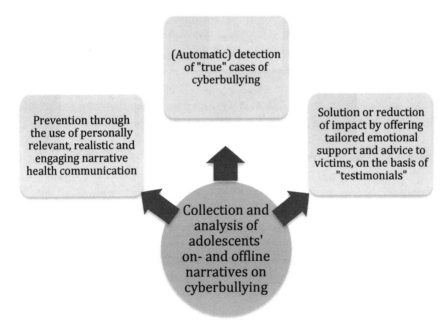

Figure 14.1. Adolescents' narratives of cyberbullying as a basis for general prevention, online detection, and response

With regard to these new technologies, it is clear that Web 2.0 characteristics such as the shareability (Papacharissi & Gibson, 2011), persistence, scalability, replicability, and searchability of digital content (boyd, 2008,

2011) are often described as problematic in the context of cyberbullying, because they allow 24/7 bullying in front of a worldwide audience. However, throughout this chapter, it became clear that the same characteristics have also created opportunities for narrative research and for health interventions. Stories can now, for instance, be gathered through interviews via chat, instant messaging, mail, or video applications, or they may even be found on video sharing sites, social networking sites, and discussion or helpline fora. New media-literate youngsters may also be invited to tell their stories through pictures or videos, using technologies such as smartphones that facilitate this type of communication. On the intervention side, persuasive (text, video, or game) narrative messages aimed at prevention may be codesigned and distributed online. Furthermore, data-mining software allows for detecting and analyzing large amounts of information (such as concrete cases of cyberbullying or stories thereof). Finally, information and communication technologies also offer methods to facilitate follow-up of detected cyberbullying cases, for instance, by automatically sending positive stories to those who have become the victim of cyberbullying. In sum, Web 2.0. seems to provide a basis for narrative methods 2.0 and for interventions 2.0, and their application in the field of cyberbullying appears to be promising.

Using ICTs in narrative research does, however, also raise many practical and ethical questions. For instance, how can we elicit, mine, record, analyze, and interpret online data? What ethical and legal issues are related to these methods, especially when minors are involved and when the topic of interest is sensitive (such as cyberbullying)? Similarly, the use of ICTs for narrative-based general prevention, detection, and solution of health-related problems poses several challenges. How can we, for instance, foster the cooperation among health professionals, creative or technological professionals, and the target group to develop appealing, effective, evidence-based tools (Bouman, 2002)? How do we balance freedom of expression, privacy rights, and self-reliance of adolescents on the one hand and (adult) control mechanisms or risk protection (such as online monitoring tools) on the other (Byron, Albury, & Evers, 2013; Van Royen et al., 2014)? Furthermore, one might question whether well-intended messages aimed at identified cyberbullying victims are always appreciated and beneficial. Some adolescents might experience the generated advice as intrusive, and the mere identification or treatment as a "cyberbullying victim" might be perceived as further stigmatizing.

As is the case for individual users of ICTs, researchers and other health intervention developers should thus try to find the right balance between what is technologically feasible and socially desirable.

References

Atkinson, R. (1998). *The life story interview*. Thousand Oaks, CA: Sage.

Bandura, A. (1977). *Social learning theory*. Upper Saddle River, NJ: Prentice Hall.

Bastiaensens, S., Vandebosch, H., Poels, K., Van Cleemput, K., DeSmet, A., & De Bourdeaudhuij, I. (2014). Cyberbullying on social network sites. An experimental study into bystanders' behavioural intentions to help the victim or reinforce the bully. *Computers in Human Behavior, 31*(0), 259–271.

Bauman, S., Cross, D., & Walker, J. L. (Eds.). (2013). *Principles of cyberbullying research: Definitions, measures, and methodology*. New York, NY: Routledge.

Bouman, M. P. A. (2002) Turtles and peacocks: Collaboration in entertainment-education television. *Communication Theory, 12*(2), 225–244.

Bowler, L., Mattern, E., Knobel, C., & Keilty, P. (2014). *Exploring cyberbullying through visual narratives*. Paper presented at the iConference 2013: Data, Innovation, Wisdom, Fort Worth, TX. Retrieved January 31, 2015 from https://www.ideals.illinois.edu/bitstream/handle/2142/42068/360.pdf?sequence=2

boyd, d. (2008). *Taken out of context: American teen sociality in networked publics* (Doctoral dissertation). University of California, Berkeley. Retrieved January 31, 2015 from http://www.danah.org/papers/TakenOutOfContext.pdf

boyd, d. (2011). Social network sites as networked publics: Affordances, dynamics, and implications. In Z. Papacharissi (Ed.), *Networked self: Identity, community, and culture on social network sites* (pp. 39–58). New York, NY: Routledge

Bülow, P. H. (2004). Sharing experiences of contested illness by storytelling. *Discourse & Society, 15*(1), 33–53.

Byron, P., Albury, K., & Evers, C. (2013). "It would be weird to have that on Facebook": Young people's use of social media and the risk of sharing sexual health information. *Reproductive Health Matters, 21*(41), 35–44.

Connelly, F. M., & Clandinin, D. J. (1990). Stories of experience and narrative inquiry. *Educational Researcher, 19*(5), 2–14.

Del Rey, R., Casas, J. A., & Ortega, R. (2012). The ConRed Program, an evidence-based practice. *Comunicar, 20*(39), 129–137.

Dempsey, A. G., Sulkowski, M. L., Nichols, R., & Storch, E. A. (2009). Differences between peer victimization in cyber and physical settings and associated psychosocial adjustment in early adolescence. *Psychology in the Schools, 46*(10), 962–972.

DeSmet, A., Veldeman, C., Poels, K., Bastiaensens, S., Van Cleemput, K., Vandebosch, H., & De Bourdeaudhuij, I. (2013). Determinants of self-reported bystander behavior in cyberbullying incidents amongst adolescents. *Cyberpsychology, Behavior, and Social Networking, 17*(4), 207–215.

Doyle, S. (2011, March). *A program effects case study of the CyberSmart! student curriculum in a private school in Florida* (master's thesis). Memorial University, St. John's, Newfoundland and Labrador, Canada.

Gámez-Guadix, M., Orue, I., Smith, P. K., & Calvete, E. (2013). Longitudinal and reciprocal relations of cyberbullying with depression, substance use, and problematic internet use among adolescents. *Journal of Adolescent Health, 53*(4), 446–452.

Green, L., Brady, D., Ólafsson, K., Hartley, J., & Lumby, C. (2011). Risks and safety for Australian children on the internet. *Cultural Science, 4*(1). Retrieved January 31, 2015 from https://www.ecu.edu.au/__data/assets/pdf_file/0009/294813/U-Kids-Online-Survey.pdf

Green, M. C., & Brock, T. C. (2000). The role of transportation in the persuasiveness of public narratives. *Journal of Personality and Social Psychology, 79*(5), 701–721.

Green, M. C., Brock, T. C., & Kaufman, G. F. (2004). Understanding media enjoyment: The role of transportation into narrative worlds. *Communication Theory, 14*(4), 311–327.

Green, M. C., & Jenkins, K. M. (2014). Interactive narratives: Processes and outcomes in user-directed stories. *Journal of Communication, 64*(3), 479–500.

Gubrium, A., & Harper, K. (2013). *Participatory visual & digital methods.* Walnut Creek, CA: Left Coast Press.

Hasebrink, U., Livingstone, S., & Haddon, L. (2008). *Comparing children's online opportunities and risks across Europe: Cross-national comparisons for EU kids online.* London, UK: EU kids online.

Heilferty, C. M. (2011). Ethical considerations in the study of online illness narratives: A qualitative review. *Journal of Advanced Nursing, 67*(5), 945–953.

Heirman, W., & Walrave, M. (2008). Pesten in bits & bytes. In B. Martens, G. Dierick, & W. Noot (Eds.), *Ethiek en weerbaarheid in de informatiesamenleving* (pp. 68–72). Leuven, Belgium: LannooCampus.

Hinduja, S., & Patchin, J. W. (2008). Personal information of adolescents on the Internet: A quantitative content analysis of MySpace. *Journal of Adolescence, 31*(1), 125–146.

Hinyard, L. J., & Kreuter, M. W. (2007). Using narrative communication as a tool for health behavior change: A conceptual, theoretical, and empirical overview. *Health Education & Behavior, 34*(5), 777–792.

Hookway, N. (2008). "Entering the blogosphere": Some strategies for using blogs in social research. *Qualitative Research, 8*(1), 91–113.

Hopfer, S., & Clippard, J. R. (2011). College women's HPV vaccine decision narratives. *Qualitative Health Research, 21*(2), 262–277.

Jacobs, N. C. L., Völlink, T., Dehue, F., & Lechner, L. (2014). Online pestkoppenstoppen: Systematic and theory-based development of a web-based tailored intervention for adolescent cyberbully victims to combat and prevent cyberbullying. *BMC Public Health, 14*(396).

Jewitt, C. (2012). *An introduction to using video for research* (Unpublished working paper). National Centre for Research Methods, Southampton, UK. Retrieved January 31, 2015 from http://eprints.ncrm.ac.uk/2259/

Jones, S., Millermaier, S., Goya-Martinez, M., & Schuler, J. (2008). Whose space is MySpace? A content analysis of MySpace profiles. *First Monday, 13*(9). Retrieved January 31, 2015 from http://firstmonday.org/ojs/index.php/fm/article/view/2202

Juvonen, J., & Gross, E. F. (2008). Extending the school grounds? Bullying experiences in cyberspace. *Journal of School Health, 78*(9), 496–505.

Kowalski, R. M., & Fedina, C. (2011). Cyber bullying in ADHD and Asperger syndrome populations. *Research in Autism Spectrum Disorders, 5*(3), 1201–1208.

Kowalski, R. M., Giumetti, G. W., Schroeder, A. N., & Lattanner, M. R. (2014). Bullying in the digital age: A critical review and meta-analysis of cyberbullying research among youth. *Psychological Bulletin, 140*(4), 1073–1137.

Kowalski, R. M., & Limber, S. P. (2013). Psychological, physical, and academic correlates of cyberbullying and traditional bullying. *Journal of Adolescent Health, 53*(1, Supplement), S13–S20.

Kreuter, M. W., Green, M. C., Cappella, J. N., Slater, M. D., Wise, M. E., Storey, D., … Woolley, S. (2007). Narrative communication in cancer prevention and control: A framework to guide research and application. *Annals of Behavioral Medicine: A Publication of the Society of Behavioral Medicine, 33*(3), 221–235.

Li, Q., & Fung, T. (2012). Predicting student behaviors. Cyberbullies, cybervictims, and bystanders. In Q. Li, D. Cross, & P. K. Smith (Eds.), *Cyberbullying in the global playground. Research from international perspectives* (pp. 99–114). West Sussex, UK: Wiley-Blackwell.

Macbeth, J., Adeyema, H., Lieberman, H., & Fry, C. (2013). *Script-based story matching for cyberbullying prevention.* New York, NY: ACM Press.

Machmutow, K., Perren, S., Sticca, F., & Alsaker, F. D. (2012). Peer victimisation and depressive symptoms: Can specific coping strategies buffer the negative impact of cybervictimisation? *Emotional and Behavioral Difficulties, 17*(3–4), 403–420.

Madden, M., Lenhart, A., Duggan, M., Cortesi, S., & Gasser, U. (2013). Teens and technology 2013. *Pew Internet & American Life Project and Berkman Society for Internet & Society.* Retrieved January 31, 2015 from http://www.pewinternet.org/2013/03/13/teens-and-technology-2013/

Madell, D. E., & Muncer, S. J. (2007). Control over social interactions: An important reason for young people's use of the internet and mobile phones for communication? *CyberPsychology & Behavior, 10*(1), 137–140.

Mann, C., & Stewart, F. (2000). Practicalities of using CMC. In *Internet communication and qualitative research* (pp. 17–38). London, UK: Sage.

Manuvinakurike, R., Barry, B., & Bickmore, T. (2013). *Indexing stories for conversational health interventions.* Proceedings of AAAI Spring Symposium: Data Driven Wellness. 2013. Retrieved January 31, 2015 from http://relationalagents.com/publications/AAAI13-ramesh.pdf

Marwick, A., & boyd, d. (2011). The drama! Teen conflict, gossip, and bullying in networked publics. In *A decade in internet time: Symposium on the dynamics of the internet and society* (p. 25). Oxford, UK: Oxford Internet Institute. Retrieved April 28, 2015 from http://papers.ssrn.com/sol3/papers.cfm?abstract_id=1926349

Marwick, A., & boyd, d. (2014). "It's just drama": Teen perspectives on conflict and aggression in a networked era. *Journal of Youth Studies, 17*(9) 1–18. Advance online publication.

Menesini, E., Nocentini, A., & Calussi, P. (2011). The measurement of cyberbullying: Dimensional structure and relative item severity and discrimination. *Cyberpsychology, Behavior, and Social Networking, 14*(5), 267–274.

Menesini, E., Nocentini, A., Palladino, B. E., Frisén, A., Berne, S., Ortega-Ruiz, R., … Smith, P. K. (2012). Cyberbullying definition among adolescents: A comparison across six

European countries. *Cyberpsychology, Behavior, and Social Networking, 15*(9), 455–463.

Miller-Day, M., & Hecht, M. L. (2013). Narrative means to preventative ends: A narrative engagement framework for designing prevention interventions. *Health Communication, 28*(7), 657–670.

Mishna, F., Khoury-Kassabri, M., Gadalla, T., & Daciuk, J. (2012). Risk factors for involvement in cyber bullying: Victims, bullies and bully-victims. *Children and Youth Services Review, 34*(1), 63–70.

Moen, T. (2006). Reflections on the narrative research approach. *International Journal of Qualitative Methods, 5*54, 56–69.

Olweus, D. (1993). *Bullying at school: What we know and what we can do.* Malden, MA: Blackwell.

Pabian, S., & Vandebosch, H. (2014). Using the theory of planned behaviour to understand cyberbullying: The importance of beliefs for developing interventions. *European Journal of Developmental Psychology, 11*(4), 463–477.

Pace, S. (2008). YouTube: An opportunity for consumer narrative analysis? *Qualitative Market Research: An International Journal, 11*(2), 213–226.

Palladino, B. E., Nocentini, A., & Menesini, E. (2012). Online and offline peer led models against bullying and cyberbullying. *Psicothema, 24*(4), 634–639.

Papacharissi, Z., & Gibson, P. L. (2011). Fifteen minutes of privacy: Privacy, sociality, and publicity on social network sites. In S. Trepte & L. Reinecke (Eds.), *Privacy online: Perspectives on privacy and self-disclosure in the social web* (pp. 75–91). Heidelberg, Germany: Springer.

Parker, I. (2003). Psychoanalytic narratives: Writing the self into contemporary cultural phenomena. *Narrative Inquiry, 13*(2), 301–315.

Price, D., Green, D., Spears, B., Scrimgeour, M., Barnes, A., Geer, R., & Johnson, B. (2014). A qualitative exploration of cyber-bystanders and moral engagement. *Australian Journal of Guidance and Counselling, 24*(1), 1–17.

Price, M., & Dalgleish, J. (2010). Cyberbullying: Experiences, impacts and coping strategies as described by Australian young people. *Youth Studies Australia, 29*(2), 51–59.

Pure, R. A., & Metzger, M. (2012). *The outcomes of online and offline victimization by sex: Males' and females' reactions to cyberbullying versus traditional bullying.* Paper presented at the annual meeting of the International Communication Association, Phoenix, AZ.

Pyzalski, J., & Wojtasik, L. (2010). The spotlight on electronic aggression and cyberbullying in Poland. In V. Mora-Merchán (Ed.), *Cyberbullying: A cross-national comparison* (pp. 175–188). Landau, Germany: Verlag Empirische Pädagogik.

Rathi, D., & Given, L. M. (2010). *Research 2.0: A framework for qualitative and quantitative research in Web 2.0 environments.* Paper presented at the 43rd Hawaii International Conference on System Sciences, Kauai, HI.

Riessman, C. (2006). Narrative interviewing. In V. Jupp (Ed.), *The SAGE dictionary of social research methods* (pp. 190–192). London, UK: Sage. Retrieved January 31, 2015 from http://srmo.sagepub.com/view/the-sage-dictionary-of-social-research-methods/n125.xml

Robinson, K. (2001). Unsolicited narratives from the Internet: A rich source of qualitative data. *Qualitative Health Research, 11*(5), 706–714.

Şahin, M. (2012). The relationship between the cyberbullying/cybervictimization and loneliness among adolescents. *Children and Youth Services Review, 34*(4), 834–837.

Salmivalli, C., Sainio, M., & Hodges, E. V. E. (2013). Electronic victimization: Correlates, antecedents, and consequences among elementary and middle school students. *Journal of Clinical Child & Adolescent Psychology, 42*(4), 442–453.

Salmons, J. (2009). *Online interviews in real time.* Thousand Oaks, CA: Sage.

Schneider, S. K., O'Donnell, L., Stueve, A., & Coulter, R. W. S. (2012). Cyberbullying, school bullying, and psychological distress: A regional census of high school students. *American Journal of Public Health, 102*(1), 171–177.

Schultze-Krumbholz, A., Wölfer, R., Jäkel, A., Zagorscak, P., & Scheithauer, H. (2012, June). *Effective prevention of cyberbullying in Germany—The Medienhelden Program.* Presentation at the XXth ISRA World Meeting, Luxembourg City, Luxembourg.

Šléglová, V., & Černá, A. (2011). Cyberbullying in adolescent victims: Perception and coping. *Cyberpsychology: Journal of Psychosocial Research on Cyberspace, 5*(2). Retrieved January 31, 2015 from http://cyberpsychology.eu/view.php?cisloclanku= 2011121901&article=4

Smith, P. K., Mahdavi, J., Carvalho, M., Fisher, S., Russell, S., & Tippet, N. (2008). Cyberbullying: Its nature and impact in secondary school pupils. *Journal of Child Psychology and Psychiatry, 49*(4), 376–385.

Smith, P. K., & Steffgen, G. (Eds.). (2013). *Cyberbullying through the new media: Findings from an international network.* London, UK: Psychology Press.

Snee, H. (2008). *Web 2.0 as a social science research tool* (pp. 1–34). Manchester, UK: University of Manchester. Retrieved January 31, 2015 from http://www.bl.uk/reshelp/ bldept/socsci/socint/web2/web2.pdf

Sourander, A., Brunstein Klomek, A., Ikonen, M., Lindroos, J., Luntamo, T., Koskelainen, M., . . . Helenius, H. (2010). Psychosocial risk factors associated with cyberbullying among adolescents: A population-based study. *Archives of General Psychiatry, 67*(7), 720–728.

Spears, B., Slee, P., Owens, L., & Johnson, B. (2009). Behind the scenes and screens: Insights into the human dimension of covert and cyberbullying. *Zeitschrift Für Psychologie/Journal of Psychology, 217*(4), 189–196.

Stavrositu, C., & Kim, J. (2014). All blogs are not created equal: The role of narrative formats and user-generated comments in health prevention. *Health Communication, 30*(5), 1–11.

Steffen, V. (1997). Life stories and shared experience. *Social Science & Medicine (1982), 45*(1), 99–111.

Sticca, F., Ruggieri, S., Alsaker, F., & Perren, S. (2013). Longitudinal risk factors for cyberbullying in adolescence. *Journal of Community & Applied Social Psychology, 23*(1), 52–67.

Sutherland, O., Breen, A., & Lewis, S. (2013). Discursive narrative analysis: A study of online autobiographical accounts of self-injury. *The Qualitative Report, 18*(95), 1–17.

Tokunaga, R. S. (2010). Following you home from school: A critical review and synthesis of research on cyberbullying victimization. *Computers in Human Behavior, 26*(3), 277–287.

Valkenburg, P., & Peter, J. (2011). Online communication among adolescents: An integrated model of its attraction, opportunities, and risks. *Journal of Adolescent Health, 48*(2), 121–127.

Vandebosch, H., & Poels, K. (2012). Friendly ATTAC: Virtuele scenario's tegen cyberpesten. In F. Goossens, M. Vermande, & M. van der Meulen (Eds.), *Pesten op school. Achtergronden en interventies* (Bullying at school. Background and interventions) (pp. 181–186). Den Haag, Netherlands: Boom Lemma.

Vandebosch, H., & Van Cleemput, K. (2008a, May). *What is cyber bullying?* Paper presented at the International Communication Association Conference, Montreal, Canada.

Vandebosch, H., & Van Cleemput, K. (2008b). Defining cyberbullying: A qualitative research into the perceptions of youngsters. *CyberPsychology & Behavior, 11*(4), 499–503.

Vandebosch, H., & Van Cleemput, K. (2009). Cyberbullying among youngsters: Profiles of bullies and victims. *New Media & Society, 11*(8), 1349–1371.

Vandebosch, H., Van Cleemput, K., Mortelmans, D., & Walrave, M. (2006). *Cyberpesten bij jongeren in Vlaanderen, studie in opdracht van het.* (Cyberbullying among youngsters in Flanders.) Retrieved February 9, 2015 from http://wise.vub.ac.be/fattac/mios/Eindrapport%20cyberpesten%20viwta%202006.pdf

Van Royen, K., Poels, K., Daelemans, W., & Vandebosch, H. (2014). Automatic monitoring of cyberbullying on social networking sites: From technological feasibility to desirability. *Telematics and Informatics, 32*(1), 89–97.

Varjas, K., Talley, J., Meyers, J., Parris, L., & Cutts, H. (2010). High school students' perceptions of motivations for cyberbullying: An exploratory study. *Western Journal of Emergency Medicine, 11*(3), 269–273.

Walton, G., & Niblett, B. (2013). Investigating the problem of bullying through photo elicitation. *Journal of Youth Studies, 16*(5), 646–662.

Williams, K. D., & Jarvis, B. (2006). Cyberball: A program for use in research on interpersonal ostracism and acceptance. *Behavior Research Methods, 38*(1), 174–180.

Willox, A. C., Harper, S. L., & Edge, V. L. (2013). Storytelling in a digital age: Digital storytelling as an emerging narrative method for preserving and promoting indigenous oral wisdom. *Qualitative Research, 13*(2), 127–147.

Produsing Ethics
[for the Digital Near Future]

Annette N. Markham

At the close of 2014, ethics seemed heavy on the public mind. Public attention certainly swelled when Edward Snowden revealed in 2013 that the National Security Agency was collecting massive amounts of data on its citizens. Then, in midsummer 2014, the public learned Facebook had been conducting experiments on its users without their knowledge.[1] Concerns about hidden behavioral research and oversight only contributed to existing distress about data privacy. Shortly thereafter, a clash among game developers exploded into rampant denigration of and physical threats toward female game developers through social media. What was called "Gamergate" added a different set of ethics issues to the conversation—the ethics of hate speech and harassment. Although the misogynistic harassment of all feminists everywhere abated somewhat by the start of 2015, the controversy continues.

How do we respond to such situations, as individuals, scholars, designers, and policy makers? If ethics are a matter of perspective, does it matter how we respond? One might generally answer this question by saying, "Of course, it matters!" But even if one feels strongly that things seem to be heading in the wrong direction, can one change the course of history in the making, given the scope and scale of such situations? Of course, ethics is not a simple concept in these or other situations wherein values for various stakeholders clash. At the same time, it might not be that complicated, either. There is value in drawing attention to the everyday use of the concept and the continual construction of ethics through both routine and singularly unique actions: by governments, companies, designers, individuals using social media, and digital platforms themselves.

My goal in this chapter is to say emphatically that ethics matter. A more specific goal is to identify the power of the everyday produsage of ethics. This argument is grounded in the notion that "doing the right thing" is an outcome of rhetorically powerful tangles of human and nonhuman elements, embedded in deep—often invisible—structures of software, politics, and habits. Each action we take as individuals—whether designers, programmers, marketers, researchers, policy makers, or consumers—reinforces, resists, and reconfigures existing ethical boundaries for what is acceptable and just. One question emerging from recent events is "How can accountability and responsibility be more deliberately built into sociotechnical political systems of meaning making?" This question seems so huge it defies one's ability to comprehend, much less address or answer. Yet if we want to influence the shape of our digital near and distant futures, we should acknowledge that our everyday actions and statements matter: That habits and routines become, over time, taken-for-granted ways of seeing the world around us. That habits writ large become social and institutional structures. That Facebook and OkCupid and *Grand Theft Auto* are not "them," they're us, and that there-fore, we contribute to their shape and content in our everyday interactions with these platforms. That every little irritant of the Internet—whether it's being tracked by advertisers, overly reliant on apps that require giving away every piece of personally identifiable information possible, or algorithmical-ly determined to be a particular gender based on Google search habits—is at this point in time more changeable than it will ever be again, as it becomes more embedded and invisible as part of everyday practice. Unless we take the responsibility that goes along with the increasingly mundane activities of produsage, the future will seem to just happen to us. As we grow more technologically mediated and digitally saturated, it is particularly important to take ownership of the tough questions, which will allow us to develop our ethical (moral) capacities (sensibilities) to address rising issues of humanity and justice that challenge current legal and regulatory frameworks.

After a brief discussion of how our everyday sensibilities within soci-otechnical contexts are negotiated discursively, I focus on four specific examples (discourse of datafication, online quiz algorithms, academic language practices around research ethics, and public responses to ethically charged situations) to illustrate how everyday conversations, definitions, and responses can become invisible and powerful frames over time. Finally, I consider how a future-oriented ethics in practice can disrupt these frames and provoke proactive versus reactive stances.

Everyday Negotiation and Production of Ethics

When we think about ethics, we tend to think about how they precede and dictate particular actions. In academic research contexts, ethics discussions focus on how researchers treat humans in their studies. In policy making, laws and regulations—which have likely developed in particular cultural contexts over long periods of time—predetermine ethical parameters. Within professional communities, ethics tend to be operationalized through common sets of agreed-upon behaviors and attitudes. Yet, although ethics certainly shape activity, they also emerge continually from everyday activities, including the actions of software code, computational algorithms, human behaviors, design, and materiality. This social and emergent character of ethics, which is for the most part taken for granted, can be studied as a matter of everyday discourse. Formally and informally, we talk about current events, social media platforms, and various media forms, defining them in particular ways. Decisions lead to actions that over time may become habitual. As habits are absorbed and normalized, they take on obdurate and invisible properties that can direct rather than suggest future actions. These activities take place within structures that are already highly coded, templated, and normalized. Consider the discourse emerging after we learned of the Facebook contagion experiments, in which Facebook modified certain users' news feeds, displaying different types of emotional content to test the hypothesis that if users viewed happier items, they would respond in kind with happier content: sharing happier stories or posting happier updates. Reportedly, six million users were unwittingly part of this study, although the experimental intervention occurred for only a fraction of that group. This study became the topic of significant public debates about ethical research practices. One could witness a general outrage at the situation, initially focused on the hidden manipulation of users. Was Facebook legally allowed to conduct experimental research on users as if they were research subjects? As the news spread on social media, these initial questions were refined; new questions were asked. Among other things, one could notice a swiftly spreading trend to invoke the specialized and regulatory language associated with U.S. higher education research review boards (IRB, or institutional review board): Was this a "human subjects" study? Why didn't the researchers obtain "informed consent" from participants? Did Facebook or the Cornell University research team go through an IRB to get ethical approval for human subjects research? Over time, as moral and legal considerations broadened, some went further to ask: Was this a human rights violation?

Not everyone was outraged, of course. Some drew our attention to the exceedingly small possible effect of the study on individual Facebook users. Some reminded readers that if we sign an application's terms of service (TOS) agreements, we agree to a whole host of experimentation through our use of the platform/app. Others noted that because the data were anonymized, there was no need for preapproval or informed consent. Soon after this news broke, OkCupid, a social media dating service, announced proudly that it frequently conducts experiments on its users, arguing that such interventions improved the program overall.

This event exemplifies the ongoing social construction of digital technologies, whereby certain realities are up for negotiation. Although not exclusively discursive processes, the everyday talk and conversation around such events powerfully frame what will later become understood as the truth of the matter, or history. Such conversations are also additive, joining our conversations from the past 20 years as we struggle to make sense of media forms that no longer seem new but do not yet seem natural and neutral ways of being. How does this happen? Partly it's a matter of trying out different ways of making sense of anomalous or novel events. As we do this, we are making as well as using ethics.[2]

To be more specific, talking about sociotechnical structures and relations in particular ways will in turn frame how ethics are defined and practiced around and with these contexts. This process occurs at the everyday level of discourse among users, as they accept or reject certain digital platform parameters or requirements or establish certain norms for everyday use of digital and social media. It happens as people talk about their relationship with technology. It happens as people grow comfortable with certain protocols and settings in their smartphone apps and develop relations with ubiquitous technologies or the embedded Internet of things. It happens as people start to use the term "google" as a verb to mean "make a query in a browser," and it happens as people start to think of every aspect of their lives as data that can be tracked, archived, and examined as separate and discrete units of information. I agree with Silverstone (2007) that these everyday acts work at the epistemological level—not merely reflecting but making social realities.

We can point to almost any element of digital living to see this process at work. In the following sections, I sketch out four examples that cross diverse public discourses: dominant frameworks for thinking about human activity (datafication), everyday activities of consumers (online quizzes), everyday discussions of (research ethics), and public reactions and responses to ethically problematic events (e.g., the Facebook experiments or Gamergate).

Ethics Emerging From the Datafication of Everything

During a time when the idea of "big data" dominates how we think about evidence and value in Western institutions, governments, and companies, it is difficult to avoid thinking of "humans (and their data) as data" (Grinter, 2013, p. 10). As noted elsewhere, the everyday discourse of digital media includes a pervasive and "particular frame whereby everything—and I mean *every* aspect of human existence—is transformed and equalized as a unit or bit of information" (Markham, 2014). Of course we see this in advertisements for smartphones and self-tracking devices. But this frame goes beyond advertising into broader contexts whereby human experience may be understood as a complex entanglement but is simultaneously reduced and simplified into discrete, collectable, and manageable data points. This widespread tendency is particularly disturbing when applied to sensitive or precarious human situations. For instance, Global Pulse (a United Nations subcommittee addressing global crises such as poverty and epidemics) advocates real-time analysis of mobile phone, financial, and other types of digital trace data. Although not representative of its total approach to tackling humanitarian crises, Global Pulse's promotional materials reflect a reductionist datafication of humans. By framing people as "digital signals" or "well-being" as data points determined through computational analysis (Global Pulse, 2012), the complexity of human experience is diminished; the human is dehumanized.

Engaging this as a discussion of ethics means moving beyond merely acknowledging that datafication happens to consider what moralities might emerge from such a move. Even if it is possible to use computational analysis to accurately pinpoint a tipping point from "not starving" to "starving," defining humans and their experiences as data is a value-laden configuration with future consequences, not the least of which includes categorization of social groups, definition of crises, and policy making. The impact of such discourses—made with the best of intentions—may not seem immediately apparent. But the datafication frame thwarts recent efforts among policy makers and scholars to shift away from Western-centric perspectives and terminologies when talking about developing countries (Tacchi, 2012). In the case of Global Pulse, we can see a struggle between the ethic of being more sensitive to the uniqueness and participatory voices of humans in developing countries and the ethic of providing faster responses to humanitarian crises by reducing human experience to discrete and countable units of information. This is only one of many examples that demonstrate how the trend toward datafication works to construct particular ethical frameworks.

Ethics Emerging From Online Quizzes

Online quizzes provide an interesting example of ethics emerging from everyday and even banal use of technologies. Posted on sites such as Buzzfeed, these quizzes are produced and distributed as entertaining ways to determine algorithmically what Disney character we should be, what city we were meant to live in, who would be our costar in a Hollywood blockbuster, and so forth. Why do we take these? Why are they so popular? Who makes them? As Jordan Shapiro (2014) noted, it is "bizarre that any of us want to be analyzed by simple algorithms that divide and reduce us into a limited number of categories."

Such quizzes are not new, of course (they have been present on a much smaller scale in magazines for decades), and on the surface, they're quite banal. Most of us are well aware that anyone can build these fake questionnaires, with no more skill or knowledge than is required to press the "build your own survey" button. But as we take quiz after quiz, how might this repetition affect our understanding of how social and personal categories are derived? How identity is understood? How analysis of information leads to answers? Shapiro drew (2014) on the classic Freudian concept of displacement to explain our mass obsession with these quizzes, claiming that "rather than focusing on the algorithmic targeting and surveillance that has become so ordinary in our everyday lives, we distract ourselves by focusing on meaningless algorithmic categorization."

From a communication perspective, each instantiation of this "meaningless algorithmic categorization" and each calculated result utter a message at the direct level ("You are the Little Mermaid") and at an indirect or meta level ("This is how self-analysis works"). Regardless of whether we take these messages seriously, they become participants in a continual symbolic interaction process whereby our understandings of self, other, and our social worlds are coconstituted. In this sense, the technological system of medium, code, and algorithm contains moral components and functions ethically, even if we don't see it as having agency. The function of algorithms is powerful (Gillespie, 2014) but also indirect, in that we may not see the impact immediately. As Cheney-Lippold (2011) noted, this system "ultimately exercises control over us by harnessing these forces through the creation of relationships between real-world surveillance data and machines capable of making statistically relevant inferences about what that data can mean" (p. 178). Similar to Bolter (2012), Cheney-Lippold focused on the cybernetic system of continual feedback loops to describe a process whereby we lose "ownership over the meaning of the categories that constitute our identities" (2011, p. 178). In a similar fashion, I use the concept of frames or metaphors to point to the microsocial level at which technological systems function as persuasive discourse, speaking with us every time we interact with systems, influencing our everyday sense making of ourselves and the world around us. Online quizzes are just one type of system we could use to point to how ethics can emerge from the seeming triviality of everyday online activities.

Ethics Emerging From Talk About Research Ethics Themselves

Frames for the concept of ethics are in constant flux and transformation, especially in these first decades of the 21st century, when the contexts within which ethics operate (or morality becomes salient) continue to shift with converging media, digital technologies, and globally tangled networks of information flow. What constitutes ethical design of technologies, ethical use of data, and ethical research about people? To even begin to answer this question is to invoke a tangle of contingencies, definitions, and tendencies. Defining "ethical" is just one of many steps in this invocation. Generally, when discussed in the realm of scientific research, ethics is defined as a *stance* one takes, adhering to a set of values and principles about what is good or bad and therefore what actions will be right or wrong.

Whether we are talking about the ethics of interface design, ethical corporate use of data, or ethics of scientific research, two key terms remain central: "harm" and "good." The injunction to do the right thing is grounded in the more basic directives to do no harm or to maximize the good. For the most part, however, these concepts are subsumed by more direct discussions of law, regulation, procedures, norms, and common sense.

Historically, egregious harm in biomedical and psychological experiments prompted large international bodies to respond with statements about how research on humans should be conducted, such as *The Nuremberg Code,* the *Declaration of Helsinki,* and the *Belmont Report.*[3] Ethical principles emerging from these and other international efforts include respect, beneficence, and justice. These principles have transformed into widely accepted procedures regarding research of human subjects, including obtaining informed consent, ensuring protection of privacy, taking special care with participants deemed part of a predefined vulnerable population, and conducting research within an appropriate risk/benefit ratio.

These guidelines have become, over time, strong regulations of scientific research. IRBs exist at all U.S.-based institutions of higher education to review human subjects research proposals to determine whether they adequately conform to ethical guidelines—that is, whether the research meets a particular, predetermined set of ethical criteria and measures. During the past 20 to 30 years, ethnographic and social science research disciplines have challenged the foundations and criteria for IRB approvals, but for the most part, at least in the United States, the strong tendency has been to acknowledge the limitations of the current regulatory framework but not change the system.

Looking back, we can see that the two initial decades of Internet-related research mark a tipping point whereby regulatory boards finally acknowledged the serious failure of regulation-based frameworks to adequately predict or account for possible harm in digitally saturated, Internet-mediated contexts. The more informational our everyday lives become, the more we comprehend the complexity of the assemblages that are created—temporary, negotiated, informational fields of meaning, highly localized in context but globally networked in structure. In such contexts, the ethical stance best suited to the context cannot be universalized or determined a priori.

Within the broader Internet research community of practice, we can see strong trends to shift toward context-sensitive, grounded approaches.[4] Along a parallel track of scholarship, U.S.-based interpretive sociologists make increasingly strong arguments for contextual or communitarian ethical approaches in social research (e.g., Christians, 2000; Lincoln, 2005; Thomas, 2004). These, along with arguments across a range of disciplines, have been associated with helping shift national and international guidelines, to, for example, "recognize and facilitate emergent, case-by-case considerations of informed consent [and recommend] significant changes to the way informed consent is conceptualized within the complexities of twenty-first-century information and data contexts" (Markham & Buchanan, in press).

Despite the development of nuanced approaches for the operationalization of ethical concepts in scientific research, the general language surrounding ethics has remained ensconced in that of regulations, requirements, and concepts. These discourses privilege and preserve top-down approaches to ethical practice. Thus, traditional frames persist, albeit more flexibly. For example, when we seek exceptions to the requirement for informed consent, the old rule is still firmly in place—ethical research conduct *ought to* involve informing participants of the character and intent of the study in advance, so the participant has adequate comprehension and ability to consent or withdraw. The frame signifies not just a moral but an epistemological rhetoric, reinforcing an argument that the researcher is separate from the researched and that the intent and outcomes of research can be known in advance, both of which are contested premises.

Ethics Emerging From Everyday Responses to Ethically Problematic Public Events

Particular morality emerges as we talk about the ethics of certain events. Returning to the Facebook contagion study mentioned earlier, we can witness at least two types of public media response: First, the researchers failed to meet regulatory requirements by (a) neglecting to seek and gain approval from their IRB or (b) neglecting to seek or obtain informed consent. Both of these could be interpreted to imply that if such regulations had been met, we would have no cause for concern. Second, Facebook possibly caused harm by (a) manipulating our emotions and (b) deceiving millions of users. Although the issue is directly about harm, the primary question seems to be about regulations: Does a company have the right to use our data in such a way without our consent? Is this an ethical violation?

The tendency to look toward regulations and laws flavored both types of responses. We see this directly in debates about regulatory requirements. But in this case, we also saw it indirectly in how the Facebook experiments were framed or defined discursively. Not unexpectedly, they were compared to Stanley Milgram's infamous 1960s experiments[5] on obedience and authority, where participants were deliberately not informed they were the subjects of his experiment, were entrapped through continuous verbal insistence they continue the study even when they asked to withdraw, and were potentially subjected to serious psychological harm during these experiments. By analogous reasoning, one might conclude Facebook was acting unethically by manipulating participants through subterfuge. On the opposite end of the ethical spectrum, the experiments were compared to A/B testing, which, in marketing terms, is a widely accepted practice of testing the efficacy of different messages on consumers. Because most technology companies engage in A/B testing, one might conclude that Facebook was operating well within expected legal norms for corporate entities whose bottom line depends on effective marketing.

Deterding (2014) argued that what he called the "split reaction" in the framing of the Facebook study "points to the larger issue how we should frame and regulate private entities engaging in scientific research—and even more fundamentally, how to frame and regulate digital entrants to existing social fields." Although Deterding's analysis provides a nuanced response to the situation, even he does not escape the regulatory framework driving most of our discussions about ethics. This is not surprising, because the vast majority of responses to the situation reflected an emphasis on regulatory or legal issues.

There are, however, moments in this particular Facebook conversation where we move to different discussions about ethics. Tufekci (2014), for example, invoked Gramsci's model of hegemony to articulate how we are seduced into behaviors by social media. More to the point of research ethics, she strongly critiqued the way that the public and academia simply accept this social control:

> I'm struck by how this kind of power can be seen as no big deal. Large corporations exist to sell us things, and to impose their interests, and I don't understand why we as the research/academic community should just think that's totally fine, or resign to it as "the world we live in". [sic] That is the key strength of independent academia: we can speak up in spite of corporate or government interests.

Her critique attempted to shift the conversation of ethics to another front. She continued:

> It is clear that the powerful have increasingly more ways to engineer the public, and this is true for Facebook, this is true for presidential campaigns, this is true for other large actors: big corporations and governments . . . . That, to me, is a scarier and more important question than whether or not such research gets published.

boyd (2014) also attempted to shift the grounds of the debate, noting that if we focus too much on the research practices, we trivialize or ignore the larger issues. She reminded us that Facebook, through design and algorithms, curates content. Agreeing with Tufekci that these issues go well beyond research practice to questions of power, she spoke directly to how curating naturally involves ethical decision making: "This is a hard ethical choice at the crux of any decision of what content to show when you're making choices. And the reality is that Facebook is making these choices every day without oversight, transparency, or informed consent." The Facebook study, she noted, "provided ammunition for people's anger because it's so hard to talk about harm in the abstract," a point that helps us move to ethics discussions focusing on larger issues of potential harm and power, rather than remaining stuck in the somewhat simplistic arena of regulation and, more specifically, informed consent. Both Tufekci and boyd highlighted new issues that emerge when we reconsider what should be the topic of discussion. By deliberately shifting away from the narrow conceptual focus of *research* ethics, we begin to see other exigencies.

These four examples—datafication, online quiz algorithms, everyday terminology related to research ethics, and everyday responses to public ethics situations—illustrate how ethics are emergent, not a priori. I shift now to considering how we can use, rather than just accept, this everyday epistemological work as an opportunity to shape our future imaginaries more deliberately.

Future-Oriented Ethics: Produsage and Phronesis

Ethics are generally defined in the past tense and applied in advance, using predictive theories as a way of calculating the chances of potential harm or good. Generally intentional, this thinking enables us to compare what was with what might be to find logics for moving forward. This is not a bad strategy, but because it mostly functions to maintain the status quo, it may not be the best strategy for thinking about the future in terms of the ethical accountability of various biological, technological, and corporate entities.

We find ourselves at a critical juncture where we can and should participate in the work and play of disassembling and reassembling various possible frames that can guide our practical knowledge and practice of doing the right thing—what Silverstone (2007) might describe as a crucial type of media literacy. This requires more than a concerted effort to look forward as well as backward to consider what we wish to become. It also requires ethical judgment in practice—what Aristotle called *phronesis*. Ess (1996) and others have used this notion of phronesis in digital ethics discussions to emphasize how ethics are situated in domains of practice. One's ethical decision making always occurs in a specific context—thus, although we might also apply general principles, ethics still emerge with the exigencies of the situation. Extended from ethics to all forms of social knowledge, phronesis is also a way of emphasizing how our understandings of how the world works are always entangled with how these theories are used in context (Gadamer, 1960/1992). This connects closely with Bruns's (2008) concept of produsage, a useful term that emphasizes "the collaborative and continuous building and extending of existing content in pursuit of further improvement" (p. 21). Although primarily used to describe the hybrid role of user/producer in the development of content such as Wikipedia, the term produsage emphasizes the same sort of hybridity in the way everyday concepts such as ethics develop.

Future Making in Responses to Situations

In a Facebook thread about threats and harassment of women in the game design industry, media studies professor K. Jarrett (personal communication, October 15, 2014) wrote the following: "The repercussions of gamergate (and its ilk) are chilling . . . . Those of us who are educators, continuing to teach our students to be critical of all normative positions is at least one thing we can do." In calling for a proactive response, she added this: "But do we need to go further? Do we need to actively teach empowered practice? Do we teach our students how to speak back to backlash?"

Dewey (2014) also pushed readers to consider reframing the current conversation to highlight different critical issues:

> But Gamergate, crucially, isn't just about gender. It's not, contrary [to] its name, even about video games. At its heart, remember, the so-called "movement" (if an ambiguous hashtag with no leaders and no articulated goals can be called a movement), was always about how we define our shared cultural spaces, how we delineate identity, who is and is not allowed to have a voice in mainstream culture. It's about that tension between tradition and inclusion—and in that regard, Gamergate may be the perfect representation of our times.

These moments can function to shift the conversation to new grounds. By doing so, they can redefine the parameters of discussion and talk about ethics within a new territory. Multiple articulations can spark new trajectories. As hermeneutic explorations of the future, from critical standpoints, these commentaries constitute ethics in formation, even if not intentionally. Such microactions can function in the short term to raise consciousness about new or hidden elements of specific situations and in the longer term to nudge larger social systems toward different moral stances. Historically, we can witness how these processes have broad cultural impact, in examples such as the decades-long fight for and gradual normalizing of gender-neutral pronouns in the English language or the longstanding strategy on both sides of the abortion debate to name it "prolife" or "prochoice," to rhetorically influence moral attitudes and policy making.

Future Making in Conceptualizing Research Practice

Produsing ethics comes with responsibility. It shifts the burden from the regulatory arena to the personal and makes the personal political (in the feminist sense). Deliberately highlighting the future as the aim of research, the individual can more fully work within a logic of accountability. In scientific research environments, this may not be the most comfortable position, but such a grounded ethic can gain robustness and strength through reflexive conviction that one has made the best possible choice in the circumstances.

Phronesis always associates ethics with choices, decisions, and consequent actions. One's decision might have precedent or might adhere to certain regulatory guidelines. But this may be only accidental. The core of phronesis, as McKee and Porter (2009) illustrated, is that the situational features and dynamic relations among researcher, data, phenomenon, and artifact or participant drive the decision. This decision therefore might be in direct conflict with guidelines, norms, or regulations, but because it was derived in a casuistic or case-based way, it may constitute the stronger ethical path.

The Association of Internet Researchers' 2012 guidelines for ethical decision making in Internet research align both with the phronetic approach and McKee and Porter's (2009) discussion of casuistic reasoning, but they move one step further by underscoring questions as the primary enactment of this stance. The original guidelines (Ess & the AOIR Ethics Working Committee, 2002) were revised to remove almost all injunctions or recommendations and to avoid declarative statements. Believing that decisions are always based on the researcher's answers to various questions, we framed ethics as decisions, made by researchers, at critical junctures throughout specific projects (Markham & Buchanan, 2012). This move toward questions rather than concepts marks an important shift in temporal focus from a past to a future orientation. Rather than making decisions because of what has been done before, it compels one to ask, "What might happen if I do this or that?" or "How could it be otherwise?"

Future Making Through Speculative Figurations

Shifting from a focus on what *is* to a focus on that which *is not quite yet* can be aided through what Haraway (2003) called "figurations," or tropes that "make us want to look and need to listen for surprises that get us out of inherited boxes" (p. 32). Figurations facilitate critique, in that we can explore how things might be otherwise. They help us consider what we want to become.

Although a future-oriented perspective is not prominent in social science or humanities research, it is almost taken for granted in design. For feminist design and informatics scholars such as Croon Fors (2010) and Light (2011), accountability and ethics are embedded in every aspect of the research and design process. Whether in social science research, design, or everyday media consumption and use, the specific practice of speculative figuration takes into consideration the interplay of the forward and backward gaze. One can explore the present and near or distant past from the near-future perspective or consider the future from the trajectory of the past through the present. Such exploration is not concerned with what is (or was, or will be) but with possibilities and trajectories. It does not dismiss but actively transgresses boundaries of what has already been postulated and considered known. A thought-provoking example of this is Juan Salazar's (2014) recent ethnographic documentary film set in the year 2035. He called *Nightfall at Gaia* a "future fabulation," the result of an experimental method for blending actual footage of ethnographic data collected in 2011 with imagined future trajectories to yield a film that provokes the audience to consider what might happen rather than learn what has happened. Papacharissi (2012) discussed a similar orientation through the idea of reverse linearity. In her argument, the deliberate reinvention of theories of the past constitutes a form of "theoretical existentialism," which allows us to "revisit, mimic, and even rival the past" (p. 197).

Through such nonlinear remediation—not just of theory but also of our everyday sense-making practices in digitally saturated media contexts, we may more creatively inform future imaginaries. This goes beyond simply speculating forward as individuals. Inspired by the posthumanist thinking of Barad (2007), we can consider rethinking the future as a process of also rethinking causality and agency. To recognize that any "mattering," "worlding," or "becoming" is relational and entangled and that "emergence . . . is dependent not merely on the nonlinearity of relations, but on their intra-active nature . . . . With each intra-action . . . the possibilities for change are reworked" (p. 393). This also implies, as Watts, Ehn, and Suchman (2014) noted, that any future "isn't a temporal period existing somewhere beyond the present, but an effect of discursive and material practices enacted always in the present moment" (p. xxxiii). Taken to a less heady level, we can begin to enact this sensibility by disrupting traditional linear or retrospective (indeed, Western) ways of responding and thinking about our sociotechnical contexts. We can intervene in our own paths, knowing that "not even a moment exists on its own" and that every action we take "reconfigures the world in its becoming" (Barad, 2007, pp. 394, 396).

Remix as a "What If" Approach to Produsing Ethics

As Barad (2007) wrote, there is no getting away from ethics, for the very simple reason that "mattering is an integral part of the ontology of the world in its dynamic presencing" (p. 396). The way we respond to the world around us constitutes a form of mattering. If we pay attention to this, we can see opportunities for disrupting and even playfully reconsidering our responses. To shift the world by one or two small degrees is the sort of imagining that provides science fiction authors the stuff of great stories. But these speculative fabulations function well beyond storytelling to literally shape research design, design practice, everyday attitudes toward technologies, and everyday responses and interactions in lived contexts.

Taking a "what if?" orientation is an essential element of a contemporary culture of remix (Lessig, 2008). In this sense, remix is not limited to music, mashup videos on YouTube, or Internet memes, but it is more broadly considered an epistemological stance and methodological practice of recombining existing ideas in explorative ways to produce something different (Markham, 2013; Navis, 2012). This practice is understood to be momentary, provocative, and incomplete, because the larger social practices of remix will continue to reshape and morph the existing remix. This practice operates by paying close attention to previous legacies and then asking the question "What if?" to experiment with new vocabularies, conceptual frameworks, and compositions. This requires both honoring and loosening the grip of traditions that are or have been specifically configured in particular ways, for particular moments in time that may linger but do not define the future. Remix is an intensely generative process, one through which products are offered as contributions to ongoing conversations rather than as answers to persistent questions.

A remix approach to ethics enables ideas and practices about morality, responsibility, and "doing the right thing" to function as memes, in that they are intended to move; through the input and interaction of others who encounter them, they shift and morph into other imaginings. If they become too static, they are left behind to wither and die from inattention. A remix approach can constitute creative, ethical practice. It fundamentally represents the value of a continual critical interrogation of possible trajectories for design, cultural formations, and identification of what matters. It is neither neutral nor vacuous. Thus, produsing ethics is a call to breach the frames that currently shape our trajectories, to reassess and reimagine what we want to become. It places on us the responsibility and accountability for the production and use of ethics in our everyday relations with and talk about technology. Taking responsibility for one's role in not just envisioning but making the future may seem daunting, but as Silverstone (2007) argued, this is part of how media literacy happens in contemporary times. Media literacy, he argued, is a social and political project, not just a technical or critical skillset. As we become more and more interconnected, our continual production, distribution, and consumption of media are increasingly central, not only to the "conduct of the modern world" but also "to the ways in which the world is approached and comprehended" and eventually in how "those who live in that world with us are treated with respect, dignity, and care" (p. 182). Most importantly in this process, we all play a role, whether we recognize it or not, for shaping as well as using ethics as guiding concepts in digital media technology design and use. We might not attend to how our everyday use of the concept of data can oversimplify the human condition. We might not consider how participation in seemingly trivial activities such as online quizzes can create future norms for how a society might think about personalities and categories. We might not notice how the use of particular ethics-related terminology shapes how companies or scientists think about humans in their research.

We are actively participating in the overall future-making process, whether we notice or not. Within this framework, whether we call this participation literacy, remix, mattering, speculative fabulation, produsage, everyday research, or something else, the injunction and inspiration are clear. Ethics matter. And we can use this awareness to consider not only that we have agency but also how we ought to use it more deliberately in a search for a sustainable digital future.

Notes

1. A study was conducted by a group of researchers at Cornell University (Kramer, Guillory, & Hancock, 2014), using data collected by Facebook. The first media coverage of this study appeared in the *Cornell Chronicle* on June 10, 2014. On June 28, the majority of debate arose swiftly in mainstream and social media forms.

2. One could draw on any number of scholars to think about the issues throughout this section. To talk about the coconstruction of frames that can appear to have obdurate structures, I blend ideas from rhetorical theorist Kenneth Burke (1966), symbolic interactionist Carl Couch (1996), organizational sense-making theorist Karl Weick (1969), and sociologist Erving Goffman (1974). To talk about using and making ethics simultaneously, I draw on the work of feminist posthumanists Donna Haraway (e.g., 2003, 2007) and Karen Barad (2007).

3. Information and historical documentation of these and other international efforts can be found in many locations online, including the Office of Human Research Protections (OHRP), a division of the U.S. department of Health & Human Services (http://www.hhs.gov/ohrp/archive/index2.html).

4. For overviews of these trends, see Ess and the AOIR Ethics Working Committee, 2002; Markham and Buchanan, 2012.

5. Basic overviews of the studies and subsequent ethical concerns can be found in many sources, such as Wikipedia (http://en.wikipedia.org/wiki/Milgram_experiment) or Simply Psychology (http://www.simplypsychology.org/milgram.html).

References

Barad, K. (2007). *Meeting the universe halfway: Quantum physics and the entanglement of matter and meaning.* Durham, NC: Duke University Press.

Bolter, J. (2012). Procedure and performance in an era of digital media. In R. A. Lind (Ed.), *Produsing theory in a digital world: The intersection of audiences and production in contemporary theory* (pp. 33–49). New York, NY: Peter Lang.

boyd, d. (2014, July 1). What does the Facebook experiment teach us? *Social Media Collective Research Blog.* Retrieved February 28, 2015 from http://socialmediacollective. org/2014/07/01/facebook-experiment/

Bruns, A. (2008). *Blogs, Wikipedia, Second Life, and beyond: From production to produsage.* New York, NY: Peter Lang.

Burke, K. (1966). *Language as symbolic interaction: Essays on life, literature, and method.* Berkeley: University of California Press.

Cheney-Lippold, J. (2011). A new algorithmic identity: Soft biopolitics and the modulation of control. *Theory, Culture & Society, 28*(6), 164–181.

Christians, C. (2000). Ethics and politics in qualitative research. In N. Denzin & Y. Lincoln (Eds.), *Handbook of qualitative research* (2nd ed., pp. 133–155). Thousand Oaks, CA: Sage.

Couch, C. J. (1996). *Information technologies and social orders.* New York, NY: Aldine Press.

Croon Fors, A. (2010). Accountability in design. In E. Pirjo, J. Sefyrin, & C. Björkman (Eds.), *Travelling thoughtfulness: Feminist technoscience stories* (pp. 293–303). Umeå, Sweden: University of Umeå Press.

Deterding, S. (2014, June 29). *Frame clashes, or: Why the Facebook emotion experiment stirs such mixed emotion* [Blog post]. Retrieved April 26, 2015 from http://codingconduct. tumblr.com/post/90242838320/frame-clashes-or-why-the-facebook-emotion

Dewey, C. (2014, October 14). The only guide to gamergate you will ever need to read. *The Washington Post.* Retrieved February 28, 2015 from http://www.washingtonpost.com/news/the-intersect/wp/2014/10/14/the-only-guide-to-gamergate-you-will-ever-need-to-read/

Ess, C. (Ed.). (1996). *Philosophical perspectives on computer-mediated communication.* Albany, NY: State University of New York Press.

Ess, C., & the AOIR Ethics Working Committee. (2002). *Ethical decision-making and Internet research: Recommendations from the AOIR ethics working committee.* Retrieved February 28, 2015 from http://www.aoir.org/reports/ethics.pdf

Gadamer, H. G. (1992). *Truth and method* (2nd ed., J. Weinsheimer & D. G. Marshall, Trans.). Churchill, NY: Crossroad Press. (Original work published 1960)

Gillespie, T. (2014) The relevance of algorithms. In T. Gillespie, P. J. Boczkowski, & K. A. Foo (Eds). *Media technologies: Essays on communication, materiality, and society* (pp. 167-194). Cambridge, MA: The MIT Press.

Global Pulse. (2012). *An animated introduction to the UN's global pulse initiative* [Video]. Retrieved February 28, 2015 from http://www.unglobalpulse.org/about-new

Goffman, E. (1974). *Frame analysis.* New York, NY: Harper & Row.

Grinter, B. (2013). A big data confession. *Interactions, 20*(4), 10–11.

Haraway, D. (2003). *The companion species manifesto: Dogs, people, and significant otherness.* Chicago, IL: Prickly Paradigm Press.

Haraway, D. (2007). *Speculative fabulations for technoculture's generations: Taking care of unexpected country.* Catalogue of the ARTIUM Exhibition of Patricia Piccinini's Art, Vitoria-Gasteiz (Spain), 100–107.

Kramer, A., Guillory, J., & Hancock, J. (2014). Experimental evidence of massive-scale emotional contagion through social networks. *Proceedings of the National Academy of Science USA, 111*(24), 8788–8790. Retrieved February 28, 2015 from http://www.pnas.org/content/111/24/8788.full.pdf

Lessig, L. (2008). *Remix: Making art and commerce thrive in the hybrid economy.* New York, NY: Penguin Press.

Light, A. (2011). HCI as heterodoxy: Technologies of identity and the queering of interaction with computer. *Interacting with Computers, 23*(5), 430–438.

Lincoln, Y. (2005). Institutional review boards and methodological conservatism: The challenge to and from phenomenological paradigms. In N. Denzin & Y. Lincoln (Eds.), *Handbook of qualitative research* (3rd ed., pp. 165–181). Thousand Oaks, CA: Sage.

Markham, A. (2013). Remix culture, remix methods: Reframing qualitative methods for social media contexts. In N. Denzin & M. Giardina (Eds.), *Global dimensions of qualitative inquiry* (pp. 63–81). Walnut Creek, CA: Left Coast Press.

Markham, A. (2014). Undermining data: A critical examination of a core term in scientific inquiry [Online]. *First Monday, 18*(10).

Markham, A., & Buchanan, E. (2012). *Ethical decision-making and Internet research: Recommendations from the AOIR Ethics Working Committee* (Version 2.0). Retrieved February 28, 2015 from http://www.aoir.org/reports/ethics2.pdf

Markham, A., & Buchanan, E. (in press). Ethical considerations in digital research contexts. In J. Wright (Ed.), *Encyclopedia for social and behavioral sciences.* Amsterdam, Netherlands: Elsevier Press.

McKee, H. A., & Porter, J. E. (2009). *The ethics of internet research: A rhetorical, case-based process.* New York, NY: Peter Lang.

Navis, E. (2012). *Remix theory: The aesthetics of sampling.* New York, NY: Springer.

Papacharissi, Z. (2012). Afterword: Remediating theory. In R. A. Lind (Ed.), *Produsing theory: The intersection of audiences and production in contemporary theory* (pp. 195–204). New York, NY: Peter Lang.

Salazar, J. F. (2014, August 1). *Nightfall at Gaia.* Documentary pre-screening at the biennial conference of the European Association of Social Anthropologists (EASA), Tallinn, Estonia. Retrieved February 28, 2015 from http://ics.uws.edu.au/picturingantarctica/

Shapiro, J. (2014, January). The reason personality tests go viral will blow your mind. *Forbes.* Retrieved February 28, 2015 from http://www.forbes.com/sites/jordanshapiro/2014/01/18/the-reason-personality-tests-go-viral-will-blow-your-mind/

Silverstone, R. (2007). *Media and morality: On the rise of the mediapolis.* Cambridge, UK: Polity Press.

Tacchi, J. (2012). Digital engagement: Voice and participation in development. In H. Horst &

D. Miller (Eds.), *Digital anthropology* (pp. 225–241). Oxford, UK: Berg.

Thomas, J. (2004). Reexamining the ethics of Internet research: Facing the challenge of overzealous oversight. In M. Johns, S. L. Chen, & J. Hall (Eds.), *Online social research: Methods, issues, and ethics* (pp. 187–201). New York, NY: Peter Lang.

Tufekci, Z. (2014, June 29). Facebook and engineering the public: It's not what's published (or not), but what's done. *Medium.com.* Retrieved February 28, 2015 from https:// medium.com/message/engineering-the-public-289c91390225

Watts, L., Ehn, P., & Suchman, L. (2014). Prologue. In P. Ehn, E. M. Nilsson, & R. Topgaard (Eds.), *Making futures* (pp. ix–xxxix). Cambridge, MA: The MIT Press.

Weick, K. (1969). *The social psychology of organizing.* Reading, MA: Addison-Wesley.

Afterword:
What's So New About New Media?

Dennis K. Davis

What's so new about new media? Or new media theory? This book provides a variety of answers to this question, even as it leaves other important questions for scholars to continue to pursue. The chapters offer cogent and heuristic insights into how new media are being used and the roles they play in society and pave the way for the ongoing research needed to provide more definitive answers.

As Tom Lindlof notes in his discussion of interpretive communities, that concept is not a theory in the traditional sense: "Even with a more relaxed view of theory, current conceptualizations of interpretive community fail to tell us much about such basic matters as how to define its key components (e.g., strategy), where it can and cannot be applied (the boundary conditions of a theory), or the general mechanisms by which an interpretive community arises, changes, and sustains itself." Instead, interpretive community is a theoretical concept that "provides a vocabulary and a bundle of ideas for identifying and studying scenes in which the collective activity of reading texts is prominent." Most of the material presented in this book might be regarded similarly—and as Lindlof points out, such concepts can play a crucial role in both research and theory development. These concepts provide an essential starting point for scholarship and constitute an exciting agenda for research over the decades ahead.

The New Media Revolution:
Is it Déjà Vu All Over Again?

Several of the readings in this book remind us that the rise of new media was heralded two decades ago by hopes and fears of how they might transform society or radically alter individuals. New theories were needed to provide necessary perspective on these media, because many existing theories clearly did not apply to media that allowed so much user control or agency (produsers). After 20 years of new media development, it's time to reflect on what new media are or are not doing to our society and ourselves. It is also time to assess the produser theories that are being developed and consider how they differ from older media theories.

In the 1990s, we expected that new media technologies would bring fundamental changes to a media system that had serious shortcomings. We theorized about and looked for evidence of important changes. We watched the rise of the Internet and the social media that it spawned. Surely these were media with enormous potential to produce important changes. Even though the Internet was delivered to us via the infrastructure built for telephone and television, we expected that it would provide the basis for a fundamentally different form of communication. But has it done that? Or are we simply engaging in the same forms of communication using different technologies? In what sense have these media brought about important changes for individuals and society?

My academic career was largely devoted to documenting and assessing the rise of TV as a central medium in American society. In the 1960s, TV was still a new medium—one that seemed to offer many of the same possibilities for change as the new media that followed in the 1990s. TV was clearly superior in various ways to radio, movies, and print media. Satellites were making possible almost instantaneous coast-to-coast and transatlantic TV transmissions. Scholars envisioned television as a tool that could serve and revitalize communities. Marshall McLuhan predicted the rise of a global village. It was hoped that public TV, educational TV, and community cable channels would somehow offset the rise of commercial TV and realize the potential for public service thought to be inherent in the technology. Among other possible benefits, scholars speculated that TV could humanize politics and permit more direct communication between political elites and their followers. TV news had the potential to bypass the barrier to learning posed by literacy and provide a powerful means of informing and educating all members of the public. Was a new era of democracy dawning?

By 1999, it was clear that TV had failed to achieve many of these dreams (Comstock & Scharrer, 1999). Instead, it had become a medium that largely served to entertain people. Why? Were early scholars of TV wrong about its potential? In retrospect, it's not hard to see why TV failed. Commercial radio networks quickly moved to take control of TV and shape it into an entertainment medium. Efforts to establish community and public television were marginalized. Audiences readily accepted commercial TV as an entertainment medium, and once audience expectations of TV had been cultivated and viewing habits formed, the medium ceased to have potential to do much else. Efforts to use TV as an educational tool in classrooms proved difficult because school-aged children had already formed media use habits that impeded learning. They expected that TV would entertain them and were bored by content that wasn't packaged as entertainment. By the 1970s, news broadcasts were devolving into infotainment. Research on learning from network TV news that I helped to conduct found that TV news confused and misinformed people more often than it informed them (Davis & Robinson, 1989; Neuman, Just, & Crigler, 1992). Although some journalists lamented the public's inability to learn, TV networks were unconcerned as long as news attracted viewers and advertisers. Various indicators showed that public knowledge of politics stagnated or actually declined despite ever-enlarging flows of news (Entman, 1989). Research on media effects confirmed that although TV brought changes, they were small and largely incremental. TV didn't suddenly transform the United States into a society dominated by consumption, but it reinforced existing trends. Bad effects were offset by good effects. TV was a genie that had been successfully imprisoned in a bottle. Its power had been harnessed by advertisers and networks to drive an engine reinforcing social trends and guiding the orderly consumption of products and services. Its primary purpose was to earn profits by delivering mass entertainment to a largely undiscriminating and essentially passive audience.

Will new media succeed where TV failed? Will its potential to forge new communities and educate engaged users be realized? Are we witnessing the rise of innovative uses and services that are transforming society in useful ways? Or is this new genie being imprisoned once again? A quick assessment of companies involved in providing new media technology and services is not encouraging. New media have fueled the rise of powerful new companies such as Google, Facebook, Amazon, Apple, and Microsoft. Although some old media companies are in decline, others have found ways to lessen their dependence on legacy media and use new media to generate profits. After decades of innovation and struggle, these old and new media companies appear to be consolidating their control over new media.

We may be at a tipping point. Innovative uses of new media might allow produsers more control, or over the next decade, the potential of new media could be harnessed by big media companies in much the same way that they limited the potential of TV. Alternative uses of new media could be marginalized. Entertainment uses of new media could continue to gain popularity.

Produsers could be herded toward entertaining activities. People could spend even more of their time being entertained and have less time for other activities. Involvement in and concern about communities or politics could decline even more. The selective use of media to support and reinforce existing political beliefs and prejudices could continue to rise. More and more produsers could live in isolated virtual spaces shared by virtual friends who think and act as they do.

This isn't an appealing future—it's simply more of the same. Should we be more optimistic? After all, the notion that new media can make us all produsers implies that we can take more control over media and shape media to serve our personal need and interests. The notion implies that there can be communities of users who produce important content for each other. We can turn off our TV sets and turn on networks of other people who will enable us to reach our potential as individuals while at the same time helping us shape a more humane and equitable social world. It's not hard to imagine this possibility, but just how likely is it that we can use the tools provided by new media to shape this sort of future? The chapters in this book offer some interesting answers to this question. Despite mixed evidence, most authors remain somewhat optimistic that new media can enrich people's lives and empower them to achieve personally and socially worthwhile goals. I hope they are right.

The New Media Revolution in the Newsroom: Can Produsers Find a Place There?

As a scholar who has focused on the potential for journalism to aid the development of democracy, I have watched as the rise of new media has shaken the foundations of the news industry. The free distribution of news on the Internet destroyed the business model of most news companies. It even eroded the audience, advertising, and profits of the TV news companies who were themselves distributing news for free. Advertisers abandoned newspapers and TV news and found ways to use new media to distribute their messages more effectively and at lower cost. Just as importantly, young adults stopped reading and subscribing to newspapers. As they aged, they haven't returned to newspapers. They found whatever news they wanted on the Internet or (for now, at least) on cable TV. Scholars expressed alarm about the collapse of the news industry (Pickard, Sterns, & Aaron, 2009). How would democracy survive without the critical services offered by journalism? I'm less concerned; the type of journalism being practiced by big media companies was inherently problematic (Davis & Kent, 2013). It largely served the status quo. Its news production practices were flawed. The rise of new media might lead to better forms of journalism. I like the notion that journalism as a function in society can be separated from the commercial media institutions that have controlled it over the last two centuries. We can have journalism without the printing press (Shirky, 2009). But how will this journalism function be exercised? What social institutions will support it?

New media allow produsers to play an increasingly important role in news production and distribution, but thus far their involvement has been mostly marginalized or ignored by mainstream media. The citizen journalism movement spawned many produser experiments in which interested and informed members of the public team with journalists to produce news. So far, the results are not encouraging. Most journalists see themselves as necessarily maintaining control over the work of nonprofessionals. They view citizen journalism as providing free labor—a cost-saving measure rather than a path toward a fundamentally different way of doing journalism. The rise of Twitter offers another way for produsers to be involved in journalism. Again the results are not encouraging. Twitter has broken important stories and provided eyewitness reports journalists may use in constructing news about breaking events. But it has also spawned problematic rumors and reinforced pernicious stereotypes. It has disrupted the lives of people who become the focus of scorn or ridicule. Journalists find it easy to cite examples demonstrating the need for professional assessment and editing of Twitter chatter. It's hard to see how citizen journalism will overcome the preconceptions of journalists and begin to renew and revitalize how news is made. It seems more likely that the future of big news media was presaged by the purchase of *The Washington Post* by Amazon owner Jeff Bezos. Wealthy capitalists may see the usefulness of preserving newspaper brands along with certain forms of journalistic practice.

As I read the chapters in this book, I looked for evidence that my pessimistic view of the future might be wrong. Are scholars discovering important new uses of new media that go beyond entertainment? Are large numbers of people finding ways to break old media habits and do innovative things with media? If so, how likely is it that these innovations will persist and become widespread? In particular, is new media use enhancing or eroding democracy?

New Media, Social Institutions, and the Construction of Culture

Many of the chapters directly or indirectly deal with the power of media to alter the way in which culture is constructed. One of the most important insights into the role of media since the scientific study of media began in the last century involves culture construction. Human cultures are developed and maintained through communication. As we have developed larger and increasingly complex forms of social organization, we have created social institutions that serve to reinforce and stabilize culture. These institutions enable 300 million Americans living in the United States to have a sense of shared culture. But these institutions are a mixed blessing: The stronger they are, the more effectively they use media to dominate communication and the more they constrain and limit culture. When TV developed in the 1960s, it posed profound threats to our social institutions. The medium had the potential to disrupt both public and private communication in ways that had unpredictable consequences. But it didn't. The legitimacy of social institutions wasn't undermined, while TV had been harnessed to provide a powerful stabilizing influence and—despite disruptions created by the Civil Rights Movement, the Cold War, and the Vietnam War—focused the nation on entertainment while it regulated our consumption of products.

But what about new media? How well are our social institutions coping with the rise of new media? What evidence is there that they are maintaining control over culture construction? To what extent are social institutions assimilating new media into an overall media system that they can dominate? Many of the authors in this book have presented arguments that relate to these concerns.

Napoli and Obar support my concern that we may be at a tipping point for new media. They argue that produsers are being "repassified" by the trend toward the use of mobile devices to access the Internet. As smaller mobile devices proliferate and gain popularity, the use of full-featured computers is declining. Mobile devices have many limitations that discourage production of content and encourage passive reception. Does this mean that mobile devices will limit our ability to construct culture and cede control to social institutions? This should be an important area for research.

Hobbs's discussion of produsing Twitter in the classroom similarly seems less than sanguine about Twitter as an educational tool. She had expected that students would be skilled in using Twitter and that these skills could be adapted to instructional purposes. However, students were unable to engage in live tweeting, and although they occasionally used tweets for self-advocacy and frequently complimented each other, they were rarely critical. Twitter was rarely used to connect beyond the classroom. Her discussion reminded me of the frustrations that educators encountered when they brought televisions into classrooms in the 1960s and found that students were bored and unengaged with content that wasn't packaged as entertainment. Is something similar happening with Twitter? We need research that assesses the expectations and habits of Twitter users.

Gray's exploration of the liberatory potential of Black cyberfeminism explicitly engages whether new media might help overthrow dominant cultural institutions. She outlines the potential offered by new media and advocates the use of these media by Black feminists to gain power in the social world. She points out several positive examples of new media use but also discusses disruptive ideological disputes. Racial and ethnic groups marginalized by mainstream media frames are not finding it easy to use new media to oppose these frames. If Gray's argument is correct—that if new media are going to be used as an important tool for social change, Black cyberfeminists will need to be leaders—research will be needed to guide use of this tool.

Markham may also believe that we are at an important transitional moment, when she argues that we need to "proactively take the ethics reins and consider what we might or could become." She recognizes that ethics are embedded in the everyday media use practices that help shape how we experience the social world: "Decisions lead to action that over time may become habitual." If our habitual uses of social media are problematic, we need to make changes. This isn't always easy because so much of what we do with new media quickly becomes routinized. Markham points out many issues involved with new media use that most produsers only vaguely recognize and that are all worthy of additional consideration. But what if— given that mood management is one of the most common things people do with new media—Facebook's manipulation of users' news feeds was related to a potential algorithm to enhance people's mood? The motto of entertainment media throughout history has been to give people what they want. If they want mood enhancement, give it to them—even if it means manipulating the flow of information. People may need an unfiltered flow of information to be active, responsible citizens, but that need may be in conflict with their desire to feel good. Choosing what we want over what we need is a central ethical dilemma. With TV, the dilemma was resolved by allowing the media industries to focus on entertainment. Ethical produsers may need to actively resist technological changes and media services that focus on wants rather than needs. Researchers should study this to assess trends and advise produsers.

Woodford, Goldsmith, and Bruns discuss how to use social media audience metrics as a new way to assess TV audiences. Currently, most audience measurements are used by media to sell advertising. Old media could use new media audience metrics to provide an innovative measure of their value to advertisers. If social media users are highly involved in following and interacting around a TV program, this could demonstrate the value of a program to advertisers. So far, advertisers remain committed to audience size as the only useful audience metric—and viewer involvement is unrelated to audience size. Advertisers may wonder why they should pay attention to social media use in relation to TV programs until an innovative advertiser finds a way to exploit social media metrics. Could social media metrics be used to provide insight into culture construction? How much of this culture construction involves superficial entertainment? Are there innovative forms of culture construction taking place in relation to specific programs?

Gajjala, Tetteh, and Birzescu are concerned about the way that subalterns are marketed by development agencies to donors. They provide numerous examples demonstrating that current marketing strategies frame subalterns in demeaning and stereotypical ways. Unless and until larger cultural changes occur, perhaps we cannot expect that marketing subalterns will be any different from marketing any other commodity; advertisers use frames that can be easily understood by audiences, and the development agencies have a neoliberal agenda when they seek to transition subalterns to become successful by Western standards—in part because that is how they receive funding. Is the transitioning subaltern perhaps a sign that changes may be on the way?

Gorham and Riccio explain how social media could play a critical role in the formation of adolescent identity. Children could actively shape their identity using social media, but will they, and in what ways? If Napoli and Obar are right, children will increasingly use mobile media that impose critical limitations on communication. What if tech-savvy parents and other authority figures encourage mobile media use because it inhibits teens' ability to experiment with identity in ways not supported by cultural institutions? Might current trends in new media use mean that new media will have decreased ability to affect identity formation? Future research can look at this.

In contrast to mobile media's possible passification of the audience, Bowman discusses how video game playing is evolving into an increasingly demanding activity of coproduction. Future research can help us understand whether and how the cognitive demands of this type of gameplay differ from those associated with other (including nonmediated) games and whether publicly played games (such as gaming contests) generate different cognitive demands. Importantly, Goffman (1974) argued that the frames learned while playing were important for framing serious actions. If video games are becoming more demanding and involving, does this increase their ability to teach frames? If so, what frames are being taught? Is there an association between type of frame and cognitive demand?

Vandebosch et al. detail the difficulty of defining and addressing cyber-bullying. New media have facilitated a new form of bullying but may also support mechanisms for discovering and responding to such behavior. The recommendation to involve the audience in creating anticyberbullying materials may—as more research is conducted—ultimately help us evaluate whether a universal definition of cyberbullying could/should be developed. Or should we expect some differences in bullying from one culture to another? Looking for differences across different cultures could be enlightening. Would a study of these differences indicate whether cyberbullying is likely to be more or less problematic in certain cultures? We might gain useful insights into culture construction by conducting this research.

Epstein's innovative model for studying the development of Internet regulation places government policy makers at the center of policy making and considers the entities that influence their work. As do many models based on structuration theory, this more effectively describes the status quo than how that status quo might be changed. Overall, new media policy making has trended toward reining in the power of the technology and toward supporting the financial interests of companies directly benefitting from how technology is regulated. Future research looking into the public's role in influencing policy could investigate the impetus behind public action. Epstein's example of the flood of public concern over SOPA orchestrated by Google and the Electronic Frontier Foundation (EFF) is not reassuring. It is an example of public involvement in policy making—but would the public have been so involved if not motivated by Google and the EFF? There are many motivations to engage in the process of influencing our social struc-tures, of course, and they are worthy of our attention.

The use of social media could represent a major shift in increasing awareness of human rights violations. McPherson explains how and why human rights organizations are being forced to monitor and verify the flow of information about human rights abuses worldwide. On the one hand, they can use the ever-increasing flow of information about abuses from average citizens using social media. But the verification of that information can be difficult and expensive, so some easily verified sources dominate human rights reports. Human rights groups face the same problem of verification that journalists face when they work with citizen journalists or seek to mine Twitter feeds. But in addition, human rights groups often must contend with information flows that are manipulated to serve ideological or political agendas.

Hunsinger discusses darknets—hidden networks of people on the Internet. Darknets provide a way to ignore or evade government policy making. If we could monitor increases in darknets, we might be able to gauge the level of public rejection of these policies. On another level, do darknets, with their premium on access to information, represent a potentially significant threat to cultural institutions? Or, because of the consummativities of the darknets, does that liberatory potential die on the vine? What would it take to overcome the obstacles?

Pavlik advocates the use of flow theory to study new media. Flow theory was developed by psychologists but is easily adapted to media theory purposes. The theory focuses on the deep involvement or optimal experience that some activities can provide for some people. Are people being caught up in the flow by their involvement with Facebook or Twitter? Could this explain why some people say or do problematic things on social media, losing jobs or friends? Could flow theory be used to assess the gaming coproduction described by Bowman? Could being caught up in the flow be a contributing factor in political apathy? If so, might this lead to the sort of inertia mitigating against using social media to change cultural institutions?

So What's New About New Media?

New media have introduced important changes in the everyday life of most people, especially younger ones. Older media are clearly being displaced, although not as quickly as was once thought would happen. TV and social media use can be complementary and mutually reinforcing. But apart from enhancing our entertainment experiences, new media do not serve many serious purposes for most people. If Gray is right, racial and ethnic minorities should be turning to new media for a wide range of serious purposes. They have the most to gain from serious uses. But are they? Increasingly attractive and involving video games seem more likely to dominate teen attention than do politics or community life. Markham is right in asserting that if we want to make more ethical use of new media, we will have to exert some conscious effort. Easily formed and comfortable old or new media use habits are likely to involve entertainment. News industries are being transformed, but so far journalists haven't changed much. As long as journalists don't change, the news is likely to remain the same despite changes in the way it is delivered.

And what's so new about new media theory? This book introduces many concepts and demonstrates their usefulness. A new agenda for media research is taking shape, but it doesn't differ too much from the agenda set by older theories. Should it? We're still concerned about both the good and bad things that media do. We are developing theories that help assess whether new media can and are being used for serious things while also documenting how personal life is often dominated by entertainment. As with earlier forms of media, the technology itself contains no magical solutions to serious problems. We need to continue developing theories to guide our understanding and use of these new communication tools. If there is to be a new media revolution, we will need to develop the theories and do the research necessary to lead it. The chapters in this book represent a valuable next step on that journey.

References

Comstock, G., & Scharrer, E. (1999). *Television: What's on, who's watching, and what it means.* San Diego, CA: Academic Press.

Davis, D. K., & Kent, K. (2013). Journalism ethics in a global communication era: The framing journalism perspective. *China Media Research, 9*(2), 71–82.

Davis, D. K., & Robinson, J. P. (1989). Newsflow and democratic society in an age of electronic media. In G. Comstock, (Ed.), *Public communication and behavior* (Vol. 2, pp. 60–102). New York, NY: Academic Press.

Entman, R. M. (1989). *Democracy without citizens: Media and the decay of American politics.* New York, NY: Oxford University Press.

Goffman, E. (1974). *Frame analysis: An essay on the organization of experience.* New York, NY: Harper & Row.

Neuman, W. R., Just, M. R., & Crigler, A. N. (1992). *Common knowledge: News and the construction of political meaning.* Chicago, IL: The University of Chicago Press.

Pickard, V., Sterns, J., & Aaron, C. (2009). *Saving the news: Toward a national journalism strategy.* Washington, DC: Free Press.

Shirky, C. (2009). *Newsrooms and thinking the unthinkable.* Retrieved February 19, 2015 from http://www.shirky.com/weblog/2009/03/newspapers-and-thinking-the-unthinkable

Contributors

Philippe C. G. Adam (PhD, EHESS Paris) is director of the Institute for Prevention and Social Research and a senior research fellow at the Centre for Social Research in Health, University of New South Wales in Sydney, Australia. Combining perspectives from social sciences and health promotion, his research helps us understand and address the individual and social factors that shape people's health behaviors. His work uniquely bridges the gap between research and prevention practice.

Kath Albury (PhD, UNSW) is an associate professor in the School of Arts and Media at the University of New South Wales in Sydney, Australia. Her current research focuses on young people's practices of digital self-representation and the role of user-generated media (including social networking platforms) in young people's formal and informal sexual learning.

Sara Bastiaensens is a PhD student and researcher in Communication Studies at the University of Antwerp. Within the scope of the Friendly ATTAC project (www.friendlyattac.be), she is conducting research on adolescents' bystander behavior in cases of cyberbullying.

Anca Birzescu holds a PhD in Communication Studies from Bowling Green State University (BGSU). Her research interests include postcolonial theory, feminist cultural studies, ethnic identity, gender and communication, digitally mediated communication and development, and mass media discourse in Eastern and Central Europe. Her doctoral research addressed Roma minority identity negotiation in postcommunist Romania. She has served as an adjunct faculty member at BGSU Firelands.

Nicholas David Bowman (PhD, Michigan State University) is an associate professor of Communication Studies and a research associate in the Interaction Lab (#ixlab) at West Virginia University. His research examines human interaction and communication technology, including social media and video games. He has published more than four dozen original research reports, written two dozen book chapters, and presented nearly 80 conference papers on the topic.

Axel Bruns (PhD, University of Queensland) is an Australian Research Council future fellow and a professor in the Creative Industries faculty at Queensland University of Technology (QUT) in Brisbane, Australia. He leads the QUT Social Media Research Group and is a coeditor of *Twitter and Society* (2014). His current work focuses on the study of user participation in social media spaces, such as Twitter, and the development of new research methodologies for working with large and real-time datasets from social media spaces.

Dennis K. Davis (PhD, University of Minnesota) is professor emeritus of Communications in the College of Communications at The Pennsylvania State University. His research and teaching interests include new media, media and politics, global media, research methods, and media theory. With Stanley Baran, he is the coauthor of *Mass Communication Theory: Foundations, Ferment, and Future.* In 2014, he received the Lifetime Achievement in Scholarship award from the Broadcast Education Association.

John de Wit (PhD, University of Amsterdam) is a professor and director of the Centre for Social Research in Health at the University of New South Wales in Sydney, Australia. His area of expertise is the social psychology of health and sexuality. His research interests encompass health communication, attitude and behavior change, sexual health promotion, and health promotion planning.

Stephanie Hemelryk Donald (DPhil, University of Sussex) is head of the School of the Arts and a professor of Comparative Film and Media at the University of Liverpool. She is currently working on a book examining the role of the migrant child in world cinema.

Dmitry Epstein (PhD, Cornell University) is an assistant professor of digital policy in the Department of Communication at the University of Illinois at Chicago. His research interests include Internet governance, information policy assumptions, and online civic engagement in policy making.

Radhika Gajjala (PhD, University of Pittsburgh) is a professor of Media and Communication at Bowling Green State University. She is the author of *Cyberculture and the Subaltern* (Lexington Press, 2012) and *Cyberselves: Feminist Ethnographies of South Asian Women* (Altamira, 2004). She has also coedited books, including *Cyberfeminism 2.0* (2012), *Webbing Cyberfeminist Practice* (2008), and *South Asian Technospaces* (2008).

Ben Goldsmith (PhD, University of Queensland) is a senior research fellow in the ARC Centre of Excellence for Creative Industries and Innovation, Queensland University of Technology, in Brisbane, Australia. His research interests include media policy, creative labor, international media production, and the app economy. He is currently working on a major research project analyzing the use of screen content in Australian schools and universities.

Bradley W. Gorham (PhD, University of Wisconsin–Madison) is an associate professor and chair of the Communications Department at the S. I. Newhouse School of Public Communications at Syracuse University. His research interests focus on the relationship between the media we consume and cognition about members of social groups.

Kishonna L. Gray (PhD, Arizona State University) is the director of the Critical Gaming Lab at Eastern Kentucky University and an assistant professor in Women & Gender Studies and the School of Justice Studies. Her research interests broadly intersect identity and new media, and her recent book, *Race, Gender, & Deviance in Xbox Live,* examines the reality of women and people of color within virtual gaming communities.

Renee Hobbs (EdD, Harvard University) is a professor in the Department of Communication Studies and director of the Media Education Lab at the Harrington School of Communication and Media at the University of Rhode Island. Her research interests include media literacy, digital literacy, children and media, contemporary propaganda, copyright and fair use for digital learning, and technology integration in K-12 schools.

Jeremy Hunsinger (PhD, Virginia Tech) is an assistant professor in Communication Studies at Wilfrid Laurier University in Waterloo, Ontario, Canada. His research agenda analyzes the transformations of knowledge in the modes of production in the information age. His current research project examines innovation, expertise, knowledge production, and distributions in hacklabs and hackerspaces. He is currently editing a second volume of the *International Handbook of Internet Research* with Lisbeth Klastrup and Matthew Allen and working on a volume about makerspaces with Andrew Schrock.

Rebecca Ann Lind (PhD, University of Minnesota) is an associate professor in the Department of Communication and an associate dean in the College of Liberal Arts and Sciences at the University of Illinois at Chicago. Her research interests include race, gender, class, and media; new media studies; media ethics; journalism; and media audiences. She is currently editing *Race and Gender in Electronic Media: Challenges and Opportunities* and will soon begin a new edition of *Race/Gender/Class/Media: Considering Diversity Across Audiences, Content, and Producers.*

Thomas R. Lindlof (PhD, University of Texas at Austin) is a professor of Media Arts and Studies at the University of Kentucky. His research interests include the cultural analysis of media, media audience theory and research, and interpretive methods in communication research.

Annette N. Markham (PhD, Purdue University) is an associate professor of Information Studies at Aarhus University, Denmark, and affiliate professor of digital ethics at the School of Communication at Loyola University, Chicago. Trained as an organizational communication scholar, her research focuses on identity formation in digital contexts, as well as innovative methodologies and ethics for exploring digital culture. Her work can be found in a wide range of international publications.

Ella McPherson (PhD, University of Cambridge) is a lecturer and an ESRC Future Research Leader fellow in the Department of Sociology at the University of Cambridge. She is also a research associate of Cambridge's Centre of Governance and Human Rights and a fellow at Queens' College. Her research interests include human rights reporting; social media, mass media, and democratization; NGO journalism; pluralism; and Mexican media.

Philip M. Napoli (PhD, Northwestern University) is a professor in the Department of Journalism & Media Studies in the School of Communication & Information at Rutgers University. His research interests include media institutions and media policy. His most recent book is *Audience Evolution: New Technologies and the Transformation of Media Audiences* (Columbia University Press, 2011).

Jonathan A. Obar (PhD, Penn State University) is an assistant professor of Communication and Digital Media Studies in the faculty of Social Science and Humanities at the University of Ontario Institute of Technology. He also serves as a research associate with the Quello Center for Telecommunication Management and Law at Michigan State University. His research has been published in a variety of academic journals and addresses the impact digital technologies have on civil liberties, civic engagement, and the inclusiveness of public culture.

John V. Pavlik (PhD, University of Minnesota) is a professor of Journalism and Media Studies in the School of Communication and Information at Rutgers, The State University of New Jersey. He has written widely on the impact of new technology on journalism, media, and society. His most recent book is Converging Media. He is codeveloper of the situated documentary, a form of location-based storytelling using the emerging mobile and wearable technology known as augmented reality.

Jaime R. Riccio is a doctoral candidate at the S.I. Newhouse School of Public Communications at Syracuse University. Her research interests include new media studies, digital literacy, children and media, and networked audiences.

Dinah Tetteh is a doctoral candidate in the School of Media and Communication, Bowling Green State University. Her research interests include women's health, ICT for health, ICT for development, social support, and media analysis. Her dissertation research focuses on how women's avowed identities, including age, religion, sexuality, and race, intersect to shape their experiences of ovarian cancer and social support.

Kathleen Van Royen is a PhD student in Communication Studies at the University of Antwerp. Her research interests are situated in the area of health promotion and adolescents, specifically on adolescents, social media, online sexual harassment, and software-based intervention strategies.

Heidi Vandebosch (PhD, University of Leuven) is an associate professor of Communication Studies at the University of Antwerp. She has been conducting research on cyberbullying (prevalence, profiles of bullies and victims, impact, evidence-based interventions, the role of schools, the police, and the news media) since 2005.

Anne Vermeulen is a PhD student and research and teaching assistant in Communication Studies at the University of Antwerp. Her main field of interest concerns the link among youngsters, ICT, and the sharing of (positive and negative) emotions with others.

Darryl Woodford (PhD, Queensland University of Technology) is a visiting fellow in the Creative Industries faculty at Queensland University of Technology in

Brisbane, Australia, and the cofounder of Hypometer, a consumer-facing tool for summarizing social media content. He has a background in Engineering and Game Studies, and his research interests include the methodologies of social media research and considering how social media data can be presented publicly, through the development of web and mobile applications.

Index

A

A/B testing, 255
access, as solution to inequalities, 163, 171
activism, 183, 184, 225–26
activity, enjoyment in, 99–100
Aden, R. C., 27
Adeyema, H., 235
adolescents. *See* cyberbullying; identity formation/development; youth
African Americans. *See* cyberfeminism, Black; women, Black
agency
    engaging, 4–8
    interactivity as, 108
    relationship with structures, 2–4, 45–46
    and responsibility, 5
agents
    exercise of power resources, 50
    interaction with structures and systems, 46
algorithms, 252
Anderson, J. A., 29
Angus, I., 128
antinecessitarian social theory, 4
Application Programming Interfaces (APIs), 147
asymmetrical bandwidth, 129
athletics, flow in, 93
attention, and understanding flow, 92
audience
    access of content by, 145
    and consumer-producer relationship, 14, 129
    evolution of, 11–12
    fragmentation of, 142–43
    meaning-making by, 12
    participation of before mass media, 127–28
    passification of, 128
    repassification of, 126, 130–35, 272
audience engagement, 12–13, 142, 143, 145–47, 153
    See also audience measurement
audience measurement
    bottom-up systems of, 144–45
    contextual factors in, 150–56
    and fragmentation, 143
    and sports metrics, 150–53
    uncertainties about, 141–42
    See also audience engagement; ratings data; social media-derived audience metrics
audience studies, 19–20, 21
    See also interpretive community studies
Australia, 146
autotelic personality, 99–100
avatar-as-other, 118
avatars, digital subaltern 2.0, 162

B

Baldwin, D. A., 5
Bandura, A., 81, 82
bandwidth, asymmetrical, 129
Banks, J. D., 115, 118
Barad, K., 260
Barry, B., 238

baseball analytics, 152–53
Battelle, J., 41
Baudrillard, J., 62, 63
Baughman, L., 31
Baumeister, R. F., 78
Bavelier, D., 110
Beck, C. S., 27
Benedetti, W., 112
Benkler, Y., 129
Berger, J., 166
Berger, P. L., 2, 3
Bezos, Jeff, 271
Bickmore, T., 238
Bijker, W. E., 48
Biocca, F., 102, 114
Birzescu, A., 9, 274
Black cyberfeminism. *See*
    cyberfeminism, Black
BlackLivesMatter, 184
blogs
    by Black women, 187
    participation in, 15
    queer Indian blogs, 30–31
body image, 81
Bogost, I., 113
Bolter, J., 14, 15, 16, 252
Booth, P., 14, 15
Bottero, W., 204
Bottom of the Pyramid (BoP)
    populations, 163, 164, 172
Bourdieu, P., 3, 195, 203, 205
Bowker, J., 78
Bowler, L., 234
Bowman, N. D., 13–14, 110, 115, 117,
    274, 276
boyd, d., 231, 232, 256
Braman, S., 43
brand communities, 27–28, 34–35
Brooks-Gunn, J., 79
Brophy, J. E., 179
Bruner, J., 223
Bruns, A., 1, 11, 15, 16, 257, 274
Busse, K., 33–34
Bussey, K., 82
Butler, J., 80

C

Call of Duty (game), 113
capitalism, integrated world, 61, 62, 63,
    64, 65, 70
Carrithers, M., 25
Černá, A., 232
challenge, 96, 98, 99
change, possibility of, 4
Chen, H., 94
Chen, M., 118
Chen, X., 78
Cheney-Lippold, J., 252
Chidester, D., 184
children. *See* cyberbullying; youth
Chock, M., 81
choice
    between darknets, 70
    and satisfaction, 111
Chow, R., 166
Chow-White, P., 162
Christakis, N. A., 119
civic engagement, using Twitter for, 225–
    26
civilian witnesses, 197–98, 200–203
Clarke, C., 178
class, in socialization processes, 3
Cole, J., 100
collective intelligence, 32
Collins, P. H., 181, 183, 184, 189
communication
    in policy making, 53
    understanding in social context, 81–
    82
Conkle, A., 16
connection, to larger whole, 101–2
connectivism, and learning, 215
connectivity, 161
    See also access; Kiva; M-PESA
constructivism, 7, 8, 46
consumer-producer relationship, 127, 129
consumers, BoP populations as, 172
consummativities, 57
    concept of, 62
    and darknets, 61–68
    as mechanism of power, 70
    relation to consumption, 62

consumption
  vs. content creation, 133–34
  relation to consummativities, 62
content creation, *vs.* consumption, 133–34
context, defined, 3
conversation, and flow, 98
Coplan, R., 78
corporations, 68–69, 163
Cover, R., 80
criminality, and darknets, 57–58, 59
crises, humanitarian, 251
Croon Fors, A., 259
Crossley, A., 142
Crossley, N., 204
crowdsourcing
  in marketing paradigms, 163
  and verification of information, 201–2
Csikszentmihalyi, M., 92, 93, 94, 95, 96
cultural capital, 196, 200–201, 204
culture construction, 272–76
curating, 256
Cutts, H., 232
cyberarguing, 231
cyberbullying, 10, 229–30, 275
  adolescents' perspectives on, 231–32
  bystander behavior in, 232, 237
  collecting stories of, 233–35
  defined, 230
  detecting, 237–38, 240
  interventions, 235–38, 240
narrative methods in research on, 239
  realities of, 232
  research on, 230–33
  responses to, aimed at victims, 238
  using ICTs in research on, 239–40
  victims' experience of, 232
cyberfeminism, 9, 176, 179–80, 181, 185
cyberfeminism, Black, 9, 183, 184–88, 273
cyberteasing, 231

D

Daniels, J., 186
Dant, T., 63–64

darknets, 10–11, 276
  access to knowledge on, 67
  anonymity projects on, 67
  choice between, 70
  consummativities in, 57, 58, 61–68
  described, 58–59
  economic value creation on, 59
  examples of, 59
  expertise required for, 58–59, 61
  hiddenness of, 61, 65
  histories of, 66–67
  hyperreal activities on, 60
  legitimization of existence of, 60–61
  myths about, 59–60
  and needs/desires, 63, 65–66
  popular understanding of, 57–58
  as power, 67
  predicated on prior forms, 64
  privacy/security on, 67–71
  produsing, 57
  relationship with produsers, 63
  as trustable, 58, 64
  use of, 59–60
data, use of, 255
datafication/dataization, 15, 164, 251
decision making, in gameplay, 107–9, 111, 116
Dei, G. J. S., 184
Deleuze, G., 67
democracy, 271
DeNardis, L., 46
depression, 95, 102
DeSanctis, G., 45, 55
design, 259
desires, 57, 63
DeSmet, A., 232
Deterding, S., 255
development, of youth, 78–79
  See also identity formation
development 2.0, 163
Dewey, C., 258
Dewey, J., 6, 7
digital divide, 159, 187
digital inequalities, 135
digital readiness, 135
discursive reflexivity, 46, 48
displacement, 252
divergent community readings, 26–27

Domahidi, E., 117
drama, 231–32
Dreier, O., 3–4
Duality Squared model of Internet
    governance, 11, 42, 50–53, 54–55

E

Earned Run Average (ERA), 152
economic value creation, on darknets, 59
education
    online/distance learning, 213–14,
        224
    use of social media in, 212, 272–73
        (*See also* Twitter)
    use of television in, 269, 273
Ehn, P., 260
Elavsky, C. M., 215
Electronic Frontier Foundation (EFF), 42,
    275
Ellcessor, E., 34
emergence, 32
empowerment
    appearance of, 160
    and availability of leisure time, 170–
        71
    in Black feminist theory, 183–84
    of Black women, 188
    employing technologies for, 175
    implied progress of subaltern
        toward, 167–68
    and marketing of technologies, 164
    platforms for, 164 (*See also* Kiva;
        M-PESA)
    using social media, 188
encoding-decoding theory, 28
enjoyment, and gameplay, 114
Epstein, D., 10, 11, 14, 15
ERA (Earned Run Average), 152
Erikson, E. H., 85
Ess, C., 257
ethics, 4, 8–9, 102
    continual construction of, 247
    in design process, 259
    emerging from datafication, 251

emerging from online quizzes, 252–
    53
emerging from research ethics, 253–
    54
emerging from responses to public
    ethics situations, 255–56
everyday production of, 249–56, 273
in Facebook contagion experiments,
    247, 249–50, 255
and flow, 95
future-oriented, 257–60, 262
importance of, 248, 262
power of everyday produsage of, 248
remix approach to, 261, 262
in research, 258–59
responsibility in produsing, 258
Evans, C., 215
expectations, and gender, 82
expertise, required for darknets, 58–59,
    61, 66

F

Facebook contagion experiments, 247,
    255, 273
fact-finding, 196–97
    See also verification of information
fan fiction, 33–34
fannish intertext, 33–34
Farrar, K. M., 114
Farrell, T. S. C., 6
feedback
    and flow, 98
    and identity formation, 77–78
    and identity shift, 85
    on social media, 81
    on Twitter, 218, 222–23
feminism, 9, 185
feminism, Black, 177, 188
feminisms, virtual, 176
    See also cyberfeminism;
        cyberfeminism, Black;
        technofeminism
Fesl, E., 186
fidelity, 85

fields, 9
  concept of, 195, 203–4
  focus on, 205–6
  human rights fact-finding as, 196, 201
  non-institutionalized movement-fields, 197
Final Fantasy (game), 112
Fischer, F., 44
Fish, S., 21, 24
Five Factor Model, 83–84
flow theory, 13, 276
  and attention, 92
  conditions shaping, 13, 95–101
  defined, 93
  described, 92–95
  and ethics, 95, 102
  as explanation for popularity of social media, 101–2
  and human relationships, 94
  and joy, 93
  in online activities, research on, 94–95
  potential applications of, 93–94
  and sense of time, 93
  social media's potential to disrupt, 95
  and television viewing, 94
Forester, J., 44
forums, 30–32, 34
Fowler, J. A., 119
frames, 275
Freedman, E., 14
Friendly ATTAC, 237
Fry, C., 235
future performance, prediction of, 155

G

Gajjala, R., 9, 14, 15, 161, 274
Gajjala, V., 167
gameplay
  attention and cognitive demands of, 110–11
  behavioral demands of, 114–16
  cooperative gaming, 117–18
  decision making in, 107–9, 111, 116
  demands of, 13–14, 274–75
  emotional demands of, 112–14
  and enjoyment, 111, 114
  and morality, 116
  and out-of-game habits, 115–16
  player-avatar relationships, 118
  social demands of, 117–18
  See also GamerGate; video games
GamerGate, 177, 247, 257–58
gamification, 15
Gandy, O. H., 196, 204, 205
Garland, J., 113
Gay Pride protest marches, 30–31
Gee, J., 111
gender, 80, 82–83
gender relations, and technologies, 180
Giddens, A., 4, 46, 47, 48
gift economy, 32
Glee (television series), 31–32
Global Pulse, 251
goal clarity, and flow, 97
Goffman, E., 79, 275
Goldsmith, B., 11, 274
Gonzales, A. L., 85
Gorham, B. W., 14, 274
governance, 44
government, meaning of, 44
governmentality, meaning of, 44
Gramsci, A., 159, 256
Gray, K. L., 9, 16, 273, 276
Green, C. S., 110
Grizzard, M., 113, 116
guilt reactions, in video games, 112–13

H

habitus, 3
Hall, S., 28
Hancock, J. T., 85
Haraway, D., 259
Hargittai, E., 135
harm, and research, 253–54, 255
hashtags, 219–20
Hawthorne, S., 179
health communication, 235–37

Hecht, M. L., 236
Heiberger, G., 215
Heilferty, C. M., 15
Hero Boy video, 193–94, 204
Highfield, T., 15
Hills, M., 14
Hinman, R., 132
Hobbs, P., 187
Hobbs, R., 10, 272
Hoplamazian, G. J., 82, 83
horizon, external *vs.* internal, 6
Horner, L., 131
Horrigan, J., 135
human rights
    evidence of violation of, 9 (*See also*
        human rights reporting)
    and Facebook experiment, 249
human rights reporting, 275
    and affordances of technologies,
        198, 204
    civilian witnessing in, 197–98, 200–
        203
    Hero Boy video, 193–94, 204
    See also verification of information
Human Rights Watch, 196, 198
Humphreys, L., 132
Hunsinger, J., 10–11, 14, 15, 276
Husserl, E., 5–6

I

ICC (identity, content, community)
    model, 15
ICT4D (information and communication
    technologies for development), 159,
    163, 170
    See also Kiva; M-PESA;
        technologies
ICTs (information and communication
    technologies). *See* information and
    communication technologies
identity
    and conversation, 98
    on darknets, 65
    and oppression, 182
    as performance, 79–80

    and social context of
        communication, 82
identity, content, community (ICC)
    model, 15
identity control, 32
identity formation/development
    factors in, 78–79
    and feedback, 77–78
    and fidelity, 85
    and gender, 82–83
    and Internet, 84–85
    online performative identity theory,
        77–85
    and social interaction, 78
    and social media, 76, 80–81, 274
    of youth, 76
    youths' role in, 84
identity performance, collaborative, 14
    See also online performative identity
        theory
identity shift, 85
Illich, I., 67
impression management, 79–80
inclusion
    lack of in technofeminism, 181
    in marketing paradigms, 163
India, 30–31
individualism, 69
inequalities, digital, 135
InformaCam, 202–3
information
    production of by civilian witnesses,
        197–98
    securing, 66
    verification of (*See* verification of
        information)
informationalization of race, 162–63
information and communication
    technologies (ICTs)
    facilitation of sources of information
        by, 194
    risks of using, 229
    support of amateurs in fact-finding
        process, 197
    using in research, 234, 239–40
    See also social media; technologies

information and communication
  technologies for development
  (ICT4D), 159, 163, 170
  See also Kiva; M-PESA
information forensics, 199
information governance, 44
  See also Internet governance
information policy making. *See* Internet
  governance; policy making
information seeking
  and mobile conversion, 131–33
  on Twitter, 212–13
information sources, and social media,
  194–95
information subsidies, 196, 204, 205
inquiry, as social, 7
institutional review board (IRB), 249,
  253–54, 255
institutions
  creation of, 2
  denaturalization of, 4
integrated world capitalism, 61, 62, 63,
  64, 65, 70
intellectualism, 183
intelligence, collective, 32
intentionality, 4–6, 9
interactivity, definitions of, 108
Internet
  civil liberties enacted through, 43
  cultural norms around use of, 42
  facilitation of information seeking
    by, 131–32
  and identity formation, 84–85
  liberatory potential of, 11–12
  and modalities of structuration, 54
  and participatory audience, 128
  politics of, 42–43
  prioritization of mobile access to,
    130 (*See also* mobile conversion)
  radical potential of, 16, 135–36
  and space to thwart negative
    representations, 183
  use of and psychological well-being,
    95
  See also social media; technologies
Internet governance, 11, 275
  as cultural politics, 43
  decision making in, 53

Duality Squared model of, 11, 42,
    50–53, 54–55
  and East Coast *vs.* West Coast
    tension, 41, 54
  as global politics of domination, 43
  and interests of elite, 42
  literature in, 44–45
  as politics of control, 43
  SOPA/PIPA protests, 41, 54
interpretation
  cultural influences of on strategies
    of, 26
  genres of, 24
interpretive community, 12, 267
  as actual group, 24
  brand communities, 27–28, 34–35
  concept of, 26
  contributions of concept of, 26–27,
    36
  criticism of, 24–25
  in Fish's formulation, 22
  focus on isolated text, 30
  forums as, 32
  Jensen on, 23–24
  kinds of, 24
  limitations in understanding of, 28
  as limiting factor on fandom, 33–34
  location of, 33–34
  research of, 28
  and social linkages between texts, 34
  suitability of for studying popular
    culture, 25
  suitability of for studying produsage,
    30
  text as basis for organizing, 33–34
  as theoretical concept, 26
  and transmedia, 32–33
  and use of reception study model,
    28–29
interpretive community studies, 20, 21–24
interstices, 1, 16
IRB (institutional review board), 249,
  253–54, 255
Isomursu, M., 132
Isomursu, P., 132

J

James, B., 152, 155
James, W., 92
Jansz, J., 107, 114
Jarrett, K., 257
Jensen, K. B., 23, 25
Jobs, S., 131
Joeckel, S., 116
John, O. P., 83
Johnson, B., 234
journalism, 270–71
joy, and flow, 93
Junco, R., 215

K

Kahn, A., 117
Kaikkonen, A., 133
Kanai, R., 111
Kantor, B., 111
Karnowski, V., 132
Keilty, P., 234
Kendall, M., 185
Kenya. *See* M-PESA
Kim, J. W., 81
Kim, S. J., 135
Kim, T., 102
Kinsella, E. A., 7, 8
Kiva, 9, 161, 164, 165–69
Klein, R., 179
Knobel, C., 234
Knobloch-Westerwick, S., 82, 83
knowledge, 60, 66–67
Kolko, B., 185
Kowert, R., 117
Kraut, R., 95
Kremar, M., 114
Kubey, R., 95
Kuriyan, R., 163

L

labor, free, 163
Lahire, B., 205–6
Lange, A., 115
Lange, R., 114, 115
Lankes, D., 217
Larson, R., 94
Lasica, J. D., 15
law
    and social change, 44
    as social system, 46
learning, 214, 215
    See also education
learning management systems (LMS),
    214
Lee, B. D., 15
leisure, availability of, 170–71
Leventhal, T., 79
Lewis, N., 112
Lichterman, P., 25
Lieberman, H., 235
Light, A., 259
limited capacity model for motivated
    mediated message processing
    (LM4CP), 111
Lind, R. A., 1
Lindley, C. A., 110
Lindlof, T., 11, 14, 267
literary theory, 21–22
LM4CP (limited capacity model for
    motivated mediated message
    processing), 111
LMS (learning management systems),
    214
Loh, K. K., 111
loneliness, and Internet use, 95
Lorde, A., 178
Luckmann, T., 2, 3
Luke, T. W., 64
Lunenfeld, P., 129, 136

M

Macbeth, J., 235
Machin, D., 25
MagiQuest (game), 14–15, 16
Mainwaring, S., 163
Malle, B. F., 5
Manuvinakurike, R., 238
marketing
    A/B testing, 255
    to BoP populations, 164
    crowdsourcing in, 163
    of Kiva, 164, 167–69
    of M-PESA, 164, 170–71
    use of ICT4D tools in, 163–64
Markham, A., 4, 8–9, 15, 273, 276
Marwick, A., 231, 232
mattering, 260
Mattern, E., 234
McDonald, K. L., 78
McGloin, R., 114
McIntyre, R., 5
McKee, H. A., 259
McLuhan, M., 268
McPherson, E., 9, 15, 275
meaning-making, 12, 21–23, 24
media, choice in consumption of, 76
media, new
    and culture construction, 272–76
    effects on democracy, 271
    expectations of, 268
    lack of use of for serious purposes,
        276–77
    potential of, 269–70
    See also Internet; social media;
        technologies; Twitter
media effects, 75, 269
media literacy, 257, 261
Meier, S., 107
mesh networks, 58–59
metadata, 198, 199, 202–3, 205
Meyers, J., 232
microlending. *See* Kiva
Milgram, S., 255
Miller-Day, M., 236
Mitra, R., 30, 31
Mittell, J., 32

mobile conversion
    and content creation *vs.*
        consumption, 133–34
    and digital inequalities, 135
    drawbacks to, 126
    and information seeking, 131–33
    and Internet's radical potential, 135–
        36
    and platform architecture, 131
    and prior experience with Internet
        access, 134–35
    and repassification of audience, 126,
        130–35, 272
    See also mobile Internet access
mobile Internet access, 125–26, 130
    See also mobile conversion
mobile media
    and suppression of radical potential
        of Internet, 16
    See also mobile conversion; mobile
        Internet access
mobile natives, 134–35
    See also mobile conversion
mobile phones
    eyewitness mode in, 205
    See also ICT; ICT4D; mobile
        conversion; mobile Internet
        access
mode 2 science production, 66, 67
Mokros, H. B., 98
mood management, 273
Moore, Will H., 202
morality, and video games, 113, 116
Morley, D., 28
Moses, L. J., 5
M-PESA, 9, 161, 164, 165, 169–71
Mueller, M. L., 43
Murphy, H., 198

N

Nacke, L. E., 110
Nafus, D., 163
Nakamura, L., 185
Napoli, P. M., 11, 14, 15, 16, 143, 272

Naroditskiy, V., 202
narrative
    control of, 175, 178
    defined, 233
    youth's use of, 80
narrative health communication, 235–37
narrative interviewing, 233–34
narrative research, 233–35
Nass, C., 114
National Security Agency, 247
"*Nationwide*" *Audience, The* (Morley),
    28
naturalistic data, collection of, 235
needs, 62–63, 65–66
Netflix, 146
networked solidarity, 201
news, production and distribution of,
    270–71
newspapers, 270, 271
Nielsen, 150
Nightfall at Gaia (film), 260
Nilan, M. S., 94
non-fields, 195, 197–98, 201
non-institutionalized movement-fields,
    197
norms
    acceptance of, 47
    and gender, 82
    reproduction of, 16

O

Obar, J. A., 11, 14, 15, 16, 272
obsessive behavior, and social media use,
    102
OkCupid, 250
Ólafsson, K., 84
Oldmeadow, J. A., 117
Oliver, M. B., 113, 115
Olweus, D., 231
Olzman, M., 16
online/distance learning, 213–14, 224
online performative identity theory, 14,
    77–86
online quizzes, 252–53

open-mindedness, 7
open network learning environment, 214
    See also Twitter
oppression
    experience of, 185–86
    and identity development, 182
    multiple, effects of, 182–83
    multiple, resistance against, 188
    ranking of, 185, 186
    recreation of on Internet, 185
order, seeking, 95–96
Orlikowski, W. J., 45, 48, 49, 50, 55
Other, subaltern as, 9–10, 164, 165–67
Other transitioning into Self, subaltern as,
    167–70
Owens, L., 234

P

Paasonen, S., 181
Paisley, W., 24
Papacharissi, Z., 8, 260
Parris, L., 232
PARs (player-avatar relationships), 118
participatory research methods, 234–35
past, recontextualizing, 181
Pavlik, J. V., 13, 15, 276
peer relationships, sensitivity to, 79
PERA (Peripheral Earned Run Average),
    153
performance
    and effects of social media, 80–81
    identity as, 79–80
Peripheral Earned Run Average (PERA),
    153
personality, 77, 83–84
personality, autotelic, 99–100
Pesce, M., 144
Pestkoppen Stoppen (website), 238
philanthropy 2.0, 162, 163
    See also Kiva; M-PESA
phronesis, 257, 258–59
Physicians for Human Rights, 196
Piglet Named Kiva, A (video), 167–68
Pinch, T., 48

PIPA (Preventing Real Online Threats to Economic Creativity and Theft of Intellectual Property Act), 41, 54
Plant, S., 180
player-avatar relationships (PARs), 118
pluralism, and human rights, 204
policy, 44, 46, 48
policy discourse, role of, 49–50
policy making
communication in, 53
duality of, 45–50
enacted through process, 48
ethics in, 249
as exercise in discursive reflexivity, 48
process of, 46–48
relationship between process and outcome of, 49
political activity, on darknets, 59–60
political imaginations, 69–70
Poole, M. S., 45, 55
poor, 163, 164, 172
Porter, J. E., 259
Postill, J., 197
power, 256
and ability to produce transformation, 183–84
consummativity as mechanism of, 70
of corporations, 69
darknets as, 67
speaking truth to, 204
power relations, in technologies, 177
pranking, 232
Preventing Real Online Threats to Economic Creativity and Theft of Intellectual Property Act (PIPA), 41, 54
privacy, 64, 65, 67–71
produsage
Bruns's concept of, 257
defined, 1–2
producers
construction of social realities, 2
idea of, 61
understanding of, 13 (*See also* audience)

professional communities, ethics in, 249
promotion
of Kiva, 164, 167–69
of M-PESA, 164, 165, 169–71
punking, 232

Q

Quandt, T., 117
quizzes, online, 252–53

R

race
feminisms' failure to address, 176
(*See also* cyberfeminism, Black)
informationalization of, 162–63
Radway, J., 22–23, 24
Rahoi, R. L., 27
ranking effects, 132
ratings data, 12, 142, 143–44
See also audience measurement
Reading the Romance (Radway), 22–23
reality, social construction of, 2
reception study model, 28–29
Reeves, B., 114
reflection, 6–7
reflective action, 7
reflective practice, 6, 7
reflexivity, 4, 8, 225–26
reflexivity, discursive, 46, 48
Relax You've Got M-PESA (video), 170
remix, 261, 262
research
ethics in, 249–50, 253–54, 258–59
and harm, 253–54, 255
narrative research, 233–35
and regulation of private entities, 255
response, focus on, 29
responsibility, in reflective action, 7
responsibility, normative, 5
reverse linearity, 260
Riccio, J. R., 14, 274
Rodman, G., 185

Rogers, R., 115
role-play, by *Glee* fans, 32
role-playing games (RPG), 112
romances, reading, 22–23
Ross, S. J., 128
Rubin, K., 78

S

Sabermetrics, 152
Safaricom. *See* M-PESA
Sakaguchi, H., 112
Salazar, J., 260
Saulauskas, M. P., 16
scaffolding, 225
Schon, D. A., 6, 7, 8
Schroder, K. C., 24
Schwartz, B., 111
science, hermeneutical, 35
Scott, D. W., 26
Scott, K. D., 187
second-screen engagement, 100, 147
security
    consummativities of, 65
    on darknets, 67–71
self
    representations of, inhibition of, 15
    sense of, 78–79
Self, subaltern as, 172
self-definition, 178, 184
self-determination, 177, 178
self-empowerment, and M-PESA, 170
self-knowledge, 184
self-representation, 169
self-schemas, 77
sense-making, 21, 24
Setrakian, L., 198
Shapiro, J., 252
Sherry, J. L., 110
Shimpach, S., 128
Silver, N., 152
Silverstone, R., 250, 257, 261
Skalski, P., 114
skill, alignment with challenge, 96, 99
Slee, P., 234

Šléglová, V., 232
smartphones. *See* information and
    communication technologies; mobile
    conversion; mobile phones;
    technologies
Smith, D. W., 5
Snowden, Edward, 247
social action, genres of, 24
social cognitive theory, 81–82, 83, 236
social contexts, 3–4
social control, acceptance of, 256
social grooming, 81
social interaction, and identity formation,
    78
social media
    analyzing posts on, 149
    and autotelic personality, 100
    and challenge, 98, 99
    and conditions for flow, 96, 101
    and connection to larger whole, 101–
        2
    as crucial source of information,
        194–95
    effects of, and personality, 77
    eyewitness mode in, 205
    and feedback, 81, 98
    and goal clarity, 97
    health consequences of, 102
    as ideal performance space, 80–81
    and identity formation, 76, 80–81,
        274
    influence of online peer
        relationships, 79
    narratives on, 80
    popularity of, 91–92, 101–2
    potential to disrupt flow state, 95
    research on use of in education, 212
    structure of networks on, 149
    use of, and personality, 83–84
    and variety, 97
    verification of information on, 195
        (*See also* verification of
        information)
    See also Internet; technologies;
        Twitter

social media-derived audience metrics,
274
    analyzing, 148–50
    contextual factors in, 153–56
    limitations of, 147–48
    persistence of ratings thinking in,
      150
    and prediction of future
      performance, 155
    Telemetrics, 152
    volumetric measurements, 150–51
social practices, contextualization of, 3–4
social systems, 46, 50
social world, possibility of changing, 4
society, policy's relationship with, 44
SOPA/PIPA protests, 41, 54
sousveillance, 68
Spasojeciv, M., 132
Spears, B., 234
Spec Ops (game), 113, 116
speculative figuration, 259
sports metrics, 150–53, 155
Srivastava, S., 83
Stein, L., 33–34
Stellmach, S., 110
Steuer, J., 102
Stop Online Piracy Act (SOPA), 41, 54
stories, creation of, 1
structuration theory, 4, 45–46
structure
    duality of, 47, 48
    relationship with agency, 2–4
    in structuration theory, 46
structures, types of, 47, 50
student debt, 213
subaltern
    appearance of empowerment of, 160
    concept of, 162
    implied progress of toward
      empowerment, 167–68
    as Other, 9–10, 164
    as Other transitioning into Self, 167–
      70
    production of, 160–61
    and production of presence, 160–61
    as Self, 170–71, 172
    shift in presentation of, 167
    in Web 2.0 context, 159

Suchman, L., 260
Super Mario Bros. (game), 108–9
surveillance, 10, 67, 68
Surveillance Camera Players (SCP), 68
Swain, C., 112
Swartz, D., 3
symbolic interactionism, 75, 79
Syria Tracker, 202
systems, as persuasive discourse, 253

T

Talley, J., 232
Tamborini, R., 110, 111, 114, 116
Tannenbaum, P. H., 95
technofeminism, 9, 180–81, 185
technologies
    abilities of to effect change, 178–79
    affordances of and human rights
      reporting, 198, 204
    and audience activity/participation,
      127
    Black women's use of, 187
    disruptive potential of, 12, 19
    and gender relations, 180
    liberatory potential of, 177
    limitations of disruptive potential of,
      136
    marketed as empowering, 164
    and participatory audience, 128
    power relations in, 177
    social construction of, 250
    See also information and
      communication technologies;
      Internet; media, new; mobile
      conversion; mobile Internet
      access; social media
technology adoption, conceptualizations
    of, 45
Telemetrics, 152
television
    as educational tool, 269, 273
    effects of, 269
    as entertainment, 268–69
    expectations of, 268
    news on, 270

threats of to social institutions, 272
viewing of, 94, 96, 100
Tetteh, D., 9, 274
text
as basis for organizing interpretive
community, 33–34
making sense of, 21–22
reading, 22–23
as resource, 29
social linkages between, 34
usage of in negotiating a reading
community, 27
writing, 22
theory
historical contingency of, 19
view of, 25–26
Thiel-Stern, S., 15, 16, 75
"Thin Line, A" (website), 235
Thompson, C., 212
time, sense of, 93
Todd, Amanda, 235
transactive memory (TMS), 117–18
transmedia, 31–33
transportation theory, 236
Tufekci, Z., 256
TweetCred, 202
Twitter
citizen journalism on, 271
feedback on, 218, 222–23
Glee fan practices on, 31–32
hashtags, 219–20
information seeking on, 212–13
learning curve in using, 216–23
in online/distance learning, 224
as pedagogical tool, 10, 215, 272–73
as resource for lifelong learning, 226
TweetCred, 202
usernames, 221
uses of, 211
using for civic engagement, 225–26
using to develop professional
relationships, 218–20
using to engage in peer social
interaction, 218
using to engage in promotion and
advocacy, 221–22
using to provide attention as reward,
222–23

using to share informational content,
220–21
using to summarize and analyze new
ideas, 217–18
Weighted Tweet Index, 153, 154

U

Unger, R. M., 4
upward mobility, 170

V

Van Cleemput, K., 231
Vandebosch, H., 10, 15, 231, 275
variety, and flow, 97
Varjas, K., 232
VCR usage, 27, 35
verification of information
and affordances of ICTs, 198
identifying original source of
information, 199
and meeting of field and non-field,
195
and metadata, 198, 199, 202–3, 205
methodologies, 196–97
resources for, 200–201
strategies of, 199–200
training initiatives for, 201
verification strategies, 195–96
verification subsidies, 196, 201–3, 204–5
Veri.ly, 202
video game controllers, 114–15
video games
GamerGate, 177, 247, 257–58
and guilt reactions, 112–13
as lean-forward medium, 114
role-playing games, 112
roots of, 108–9
as unfinished texts, 108
violence in, 112–13
See also gameplay
Von Pape, T., 132
Vygotsky, L., 214

W

Wajcman, J., 179, 180, 181
Walther, J. B., 85
Wardle, C., 199
Watts, L., 260
Weaver, A. J., 112
Web 2.0, 35
Webb, L. M., 15
Weber, R., 110, 111
Weighted Tweet Index, 153, 154
Wertsch, J., 214
Westcott-Baker, A., 111
whole-heartedness, 7
Wigand, R. T., 94
Williams, R., 92
Williams, W., 113
witnesses, civilian, 197–98, 200–203
women
    feminisms, 9, 176, 185 (*See also*
        cyberfeminism; cyberfeminism,
        Black; feminism, Black;
        technofeminism)
    harassment of in game design
        industry, 257–58 (*See also*
        GamerGate)
    lived experiences of, 186
    media portrayals of, 177–78
    power differences among, 186–87
    resistance to marginalization in
        information age, 189
    utilization of virtual technologies,
        187
women, Black
    activism of, 184
    Black feminism, 177, 188
    blogs by, 187
    and digital divide, 187
    empowerment of, 188
    extension of physical lives into
        virtual, 187–88
    imagery of, 178, 182–83
    impact of multiple oppressions on,
        182–83
    perspectives of, 181
    use of technologies, 187, 188
    See also cyberfeminism, Black

Wood, M. M., 31
Woodford, D., 11, 12, 14, 15, 274
World of Warcraft (WoW) (game), 117,
    118
Wurtzel, A., 142, 144

Y

Yep, G., 16
youth, 75–76, 84
    See also cyberbullying; identity
        formation/development
YouTube
    as crucial source of information,
        194–95
    scraping, 199

Z

zoo, 166

Digital Formations

General Editor: *Steve Jones*

Digital Formations is the best source for critical, well-written books about digital technologies and modern life. Books in the series break new ground by emphasizing multiple methodological and theoretical approaches to deeply probe the formation and reformation of lived experience as it is refracted through digital interaction. Each volume in **Digital Formations** pushes forward our understanding of the intersections, and corresponding implications, between digital technologies and everyday life. The series examines broad issues in realms such as digital culture, electronic commerce, law, politics and governance, gender, the Internet, race, art, health and medicine, and education. The series emphasizes critical studies in the context of emergent and existing digital technologies.

Other recent titles include:

Felicia Wu Song
Virtual Communities: Bowling Alone, Online Together

Edited by Sharon Kleinman
The Culture of Efficiency: Technology in Everyday Life

Edward Lee Lamoureux, Steven L. Baron, & Claire Stewart
Intellectual Property Law and Interactive Media: Free for a Fee

Edited by Adrienne Russell & Nabil Echchaibi
International Blogging: Identity, Politics and Networked Publics

Edited by Don Heider
Living Virtually: Researching New Worlds

Edited by Judith Burnett, Peter Senker & Kathy Walker
The Myths of Technology: Innovation and Inequality

Edited by Knut Lundby
Digital Storytelling, Mediatized Stories: Self-representations in New Media

Theresa M. Senft
Camgirls: Celebrity and Community in the Age of Social Networks

Edited by Chris Paterson & David Domingo
Making Online News: The Ethnography of New Media Production

To order other books in this series please contact our Customer Service Department:
(800) 770-LANG (within the US)
(212) 647-7706 (outside the US)
(212) 647-7707 FAX

To find out more about the series or browse a full list of titles, please visit our website:
WWW.PETERLANG.COM